T0318530

Handbook of Green Economics

Handbook of Green Economics

Edited by

Sevil Acar, PhD
Associate Professor
Center for Climate Change and Policy Studies
Boğaziçi University
Istanbul, Turkey

Erinç Yeldan, PhD
Professor of Economics
Economics
Bilkent University
Ankara, Turkey

Academic Press is an imprint of Elsevier
125 London Wall, London EC2Y 5AS, United Kingdom
525 B Street, Suite 1650, San Diego, CA 92101, United States
50 Hampshire Street, 5th Floor, Cambridge, MA 02139, United States
The Boulevard, Langford Lane, Kidlington, Oxford OX5 1GB, United Kingdom

Handbook of Green Economics

Notices

ISBN: 978-0-12-816635-2

Publisher: Brian Romer
Acquisition Editor: Scott J. Bentley
Editorial Project Manager: Devlin Person
Production Project Manager: Sreejith Viswanathan
Cover Designer: Miles Hitchen

Contents

Contributors

Sevil Acar, PhD
Associate Professor, Center for Climate Change and Policy Studies, School of Applied Disciplines, Boğaziçi University, Istanbul, Turkey

Izzet Ari, PhD
Middle East Technical University, Graduate School of Natural and Applied Sciences, Earth System Science Department, Universiteler Mah., Cankaya, Ankara, Turkey

Rohit Azad, PhD
Assistant Professor, Centre for Economic Studies and Planning, Jawaharlal Nehru University, New Delhi, Delhi, India

Osman Balaban
City and Regional Planning Department, Middle East Technical University, Ankara, Turkey

Shouvik Chakraborty, PhD
Research Assistant Professor, Political Economy Research Institute, University of Massachusetts, Amherst, MA, United States

Cristián Ducoing, PhD
Researcher, Department of Economic History, Lund University, Lund, Sweden

Magnus Lindmark, PhD
Professor, Centre for Environmental and Resource Economics (CERE), Economic History, Umeå University, Umeå, Sweden

Begüm Özkaynak, PhD
Professor, Department of Economics, Boğaziçi University, Istanbul, Turkey

Mark Swilling, PhD
Professor, Centre for Complex Systems in Transition, Stellenbosch University, Stellenbosch, Western Cape, South Africa

Burcu Ünüvar, MSc, PhD (ongoing)
Head of Research, Chief Economist, Head of Research at Industrial Development Bank of Turkey, Istanbul, Turkey

A. Erinç Yeldan, PhD
Professor of Economics, Economics, Bilkent University, Ankara, Turkey

Riza Fikret Yikmaz, PhD
Middle East Technical University, Graduate School of Natural and Applied Sciences, Earth System Science Department, Universiteler Mah., Cankaya, Ankara, Turkey

Introduction

Sevil Acar[1], A. Erinç Yeldan[2]

Associate Professor, Center for Climate Change and Policy Studies, School of Applied Disciplines, Boğaziçi University, Istanbul, Turkey[1], Professor of Economics, Economics, Bilkent University, Ankara, Turkey[2]

With this chapter we introduce the main components of this volume, mainly aiming at a comprehensive coverage of the analytical debates surrounding the "green economy." Organized around 10 main themes (namely conceptualization, natural capital, sustainability transitions, sectoral differentiation, smart cities, environmental justice, poverty and inequality, welfare, policy instruments, and finance), our discussions will focus on issues of sustainable development implementation failures of the past, the challenges of the present, and the opportunities of the future while understanding how concepts such as green growth, low carbon economy, circular economy, and others work together.

Background

Although various definitions of the concept of the green economy exist, we will adapt here one of the most comprehensive ones coined by the United Nations Environment Programme (UNEP). According to UNEP (2011), green economy could be defined as the *one that results in improved wellbeing and social equity, while significantly reducing environmental risks and ecological scarcities*. Besides, the report highlights that the triple crisis of financial, food, and fuel has paved the road for reconsideration of the traditional growth models. In addition, international initiatives such as the United Nations Conference on Sustainable Development (Rio+20) raised a voice for the need to a green economy transition *in the context of sustainable development and poverty eradication*.

Research on climate change has intensified on a global scale as evidence on the costs of global warming continues to accumulate. According to the Intergovernmental Panel on Climate Change (IPCC), for our planet to have a 50% chance of avoiding an undesirable rise in the global average temperature by 2°C (Celsius), concentrations of greenhouse gases ought to be stabilized at 450.....ppm (parts per million) of carbon dioxide equivalent (CO_2e). This means a total carbon budget of 870−1240 gigatons of CO_2e. Given this evidence, leaders, researchers, and civil society practitioners across the globe gathered in Paris in 2015 to seek for a renewed consensus to maintain a sustainable path of development with due respect to the rights of future generations to live healthy and prosperous lives. As the UN's Millennium Development Goals Report 2013 had already attested, these are the greatest development challenges of the 21st century.

Threat of climate change, on the other hand, is by no means a technical matter solely to be envisaged in climatic arithmetic. The mere threat itself has serious consequences directly upon the *economic realm*. A recent report by the OECD, for instance, cautions that over the next 50 years, emissions of greenhouse gases due to industrial processes and burning of fossil fuels will rise twofolds from 48,700 million tons annually to 99,500 million tons in 2060; and the consequent threat of climate change will likely reduce the rate of growth in the Asian economies by up to 5% (OECD, 2014). This will have a direct effect on the global economy as the average annual rate of growth of the world economy is projected to fall from 3.6% in 2014–30 to 2.7% over 2030–60. These gloomy projections are the end results of the likely adverse consequences related with climate change such as the rise of new bacteria and widening of respiratory diseases; decline in agricultural yields; and the social problems associated with increased scarcity of water and urban space.

Given these trends, the quest for sustainable development, made more urgent because of the uncertainties about the future climate and technology, has recently led to the realization that while economic growth has been critical in improving the standard of living of millions of people in many parts of the world, its current patterns are not only unsustainable, causing significant environmental degradation, but are also characterized by deeply inefficient production and consumption processes and management of natural resources. At the root of these problems are market and governance failures for which basic economic and regulatory instruments are available, but their systematic use as part of broader policy packages has been lacking. Part of the problem is due to the fact that development of new eco-friendly technologies typically involve positive spillovers in the form of agglomeration effects, knowledge diffusion, cross-firm externalities, and industry-wide learning; and yet, the decentralized optimization embedded in the laissez-faire actions of the markets may fail to capture these positive spillovers, and competitive equilibrium may fail to achieve the social optimum.

More importantly, decentralized laissez-faire market equilibrium based on private optimization faces the danger of *path dependence*; that is, firms may be caught up specializing in dirty technologies. Path dependence of innovation may lead firms to innovate toward maintaining dirty technologies (Aghion, 2014; Aghion et al., 2011). Firms with a history of dirty innovations tend to follow that path, creating path dependence in the long run. Thus, Aghion (2014) warns that with a narrow set of instruments, limited only to carbon taxes and energy prices, it will take a very long time for the clean innovations to catch up with the dirty technologies, and calls for complementing the carbon tax with a broader set of macroeconomic policy instruments that involve interventions toward "green technologies" as well as "green employment."

All these are happening at a time when new challenges and opportunities have emerged, including the recent and still ongoing food, fuel, and financial crises, and the growing global concern about the impact of climate change and the destruction of ecosystems and biodiversity. UN research (UN, 2013) reveals that global greenhouse gas emissions maintain their upward trend and calls for bold and decisive action. The rise in emissions has been mostly due to the fast growth in developing economies.

The 2013 UN report further notes that *the present dominant model of development is facing simultaneous multiple crises such as, depletion of natural resources and the market failures that have already marked the first decades of the current millennium.* Accordingly, this model has been ineffective in enabling a productive and decent employment market and has exacerbated the phenomenon of climate change with its various effects on the types of natural resources depletion, degradation of biodiversity, energy crisis, and food security. In contrast, the report underlines that the *green economy concept proposes to break away from the not very effective current model of development and move towards a more sustainable development paradigm that is merely characterized by low carbon emissions, rational use of resources and social inclusiveness.*

All these observations are central to *green growth*, a relatively new concept, which has captured the attention of policymakers, researchers, and civil society organizations worldwide to help design and evaluate policies that can achieve environmental sustainability efficiently, while helping to stimulate growth. This is of particular interest to fast-growing emerging market economies which are characterized by rapidly increasing environmental footprints and which seek to decouple economic growth from rising energy use and pollution generation.

In fact recent evidence reveals a burgeoning literature on the possibility of a whole set of *pro-growth environmental policies.* While it is generally understood that tighter environmental standards will be costly, Porter and van der Linde (1995) confirm, with a series of case studies, that properly designed regulation via a broad spectrum of market-based instruments such as taxes and/or cap-and-trade emissions allowances can in fact trigger innovations. This notion, later to be known as *the Porter hypothesis*, suggests that the evidence is more supportive of the "weak" version (i.e., *stricter regulation leads to more innovation*) rather than the "strong" version of the hypothesis (i.e., *stricter regulation enhances business performance—or win-win*) (Ambec et al., 2011; Brännlund and Lundgren, 2009).

Asıcı and Bünül (2012) discussed the concept of the "green economy" within a broader conceptualization of a "Green New Deal" (GND). They argued that the concept could be regarded as complementary rather than substitute to the social/ecological movements. The GND strategy was also discussed in Fitzroy et al. (2012), who argued that personal well-being and employment could in fact be raised by efficient mitigation and green fiscal policies. In the authors' words, *a 'green new deal' placing more emphasis on climate change mitigation and happiness (rather than GDP as the key proxy for welfare) could be the appropriate strategy in the current economic climate of austerity and worsening recession.* In fact, Moloney and Strengers (2014) invoked the idea of "*Going Green*" discourse and highlighted the limitations of behavior change through regulations and infrastructures involving urban form, housing, and transport. In turn, Adaman et al. (2003) emphasized the aspects of "embeddedness" as a forerunner of the argument so that to *create a sustainable organic relationship with nature, it must first be reinstituted in ways that bring it under social control.*

In the following pages of this study we intend to bring together a variety of analytical contributions to argue that developing, emerging market economies would benefit from a mix of policy instruments better targeted at the *green innovation potential*. These include not only policies to spur access to technologies and capital, but a more focused set of both supply-side "technology-push" policies (including foreign assistance to jump-start technological development) and demand-side "market-pull" policies (including price interventions and market regulations)—that should induce green innovations across many industries.

Plan of the remaining pages of this book

This study is planned in 10 related, and yet individually separate, chapters. In Chapter 1, Magnus Lindmark introduces the concept of greening within the national accounting system. Narrated over the study of historical environmental accounting for Sweden, Lindmark sets the stage for analytical conceptualization of the green macroeconomics structure.

In Chapter 2, Cristián Ducoing Ruiz deepens the accounting framework and proposes an analytical approach to handle natural capital within the context of the "green economy." He elaborates on the *weak* versus *strong* sustainability rules in regard to natural resource depletion, and the use of resource proceeds for other productive means. He provides examples for *genuine saving* estimates across a set of countries and contrasts these figures with net national saving measures.

In Chapter 3, Mark Swilling takes up the challenge of theorizing the long transitions of the global economy. He attempts to propose a new framework for understanding the complex dynamics of sustainability transitions that are already underway. In so doing, he tries to understand the dynamics of the current global polycrisis as the emergent outcome of overlaps between four dimensions of transition: sociometabolic transition, technoeconomic transition, sociotechnical transition, and long-term global development cycles. Swilling's contribution with the quest for the deep transition reveals "just" characteristics.

Chapter 4 opens the discussion on the concept of "sustainability" a la UN's Sustainable Development Goals (SDGs). Here, İzzet Arı and Rıza Fikret Yıkmaz share their reflections of main green approaches on industry and manufacturing with linkages among SDGs, the Sustainable Consumption and Production patterns, and greening of industries.

In Chapter 5, Osman Balaban introduces us with the concept of "smart" cities. In his words, cities are in a position to facilitate the transition to a greener economy as they are centers of knowledge and innovation. Balaban discusses the roles that the four key urban sectors, namely land use, buildings, transportation, and waste, can play in transitioning to a green economy and a smart city.

The next two chapters dwell upon the conceptualization of the notion of "climate justice." In Chapter 6, Begüm Özkaynak brings the issues of environmental justice, climate justice, and the green economy together. In Chapter 7, in turn, Rohit Azad and Shouvik Chakraborty tackle the climate justice concept with the aid of a global carbon tax.

Alternative welfare estimation from the perspective of green economics is the topic of Magnus Lindmark and Sevil Acar in Chapter 8. Here, the authors add an environmental dimension to societal well-being, where they account for the social costs of various emissions such as CO_2, SO_2, and NO_x for the case of the US economy. They confirm that Robert Gordon's "environmental headwinds" may definitely contribute to stagnating welfare in the future to come.

The final two chapters look at the economics of the greening design. While the instruments of greening are tackled by Erinç Yeldan in Chapter 9, the issues of global finance markets are covered in Burcu Ünüvar's contribution in Chapter 10.

References

Adaman, F., Devine, P., Ozkaynak, B., 2003. Reinstituting the economic process: (re) embedding the economy in society and nature. International Review of Sociology/Revue Internationale de Sociologie 13 (2), 357–374.

Aghion, P., 2014. Industrial Policy for Green Growth. Paper Presented at the 17th World Congress of the International Economics Association, Jordan.

Aghion, P., Boulanger, J., Cohen, E., 2011. Rethinking Industrial Policy. Bruegel Policy Brief No 04, June.

Ambec, S., Cohen, M.A., Elgie, S., Lanoie, P., 2011. The Porter Hypothesis at 20: Can Environmental Innovation Enhance Innovation and Competitiveness? Resources for the Future Discussion Paper 11-01.

Asıcı, Atıl, A., Bünül, Z., 2012. Green New Deal: a green way out of the crisis? Environmental Policy and Governance 22 (5), 295–306.

Brännlund, R., Lundgren, T., 2009. Environmental policy without costs? A review of the Porter hypothesis. International Review of Environmental and Resources Economics 3 (2), 75–117.

Fitzroy, F., Franz-Vasdeki, J., Papyrakis, E., 2012. Climate change policy and subjective well-being. Environmental Policy and Governance 22 (3), 205–216.

Moloney, S., Yolande Strengers, 2014. 'Going green'?: The limitations of behaviour change programmes as a policy response to escalating resource consumption. Environmental Policy and Governance 24 (2), 94–107.

OECD, May 2011. Towards Green Growth. OECD Pub, Paris.

OECD, 2014. Policy Challenges for the Next 50 Years. OECD Pub, Paris.

Porter, M., van der Linde, C., 1995. Toward a new conception of the environment-competitiveness relationship. Journal of Economic Perspective 9 (4), 97–118.

UNEP, 2011. Towards a Green Economy: Pathways to Sustainable Development and Poverty Eradication – A Synthesis for Policy Makers. www.unep.org/greeneconomy.

United Nations, 2013. Millennium Development Goals Report (New York).

World Bank, 2012. Inclusive Green Growth: The Pathway to Sustainable Development (Washington DC).

CHAPTER

1

Greening the national accounts: basic concepts and a case study of historical environmental accounting for Sweden

Magnus Lindmark

Introduction

This chapter presents reconstructions of historical environmental accounts for Sweden for the post-1970 period with the purpose of both presenting key methodological issues pertaining to historical environmental accounting and demonstrating how historical series can be used for analysis of long-term economic and environmental change.

Historical environmental accounting is important, since the metrics produced by official statistical bodies only cover relatively short periods. This is problematic since environmental adaptation involves long-term economic and regulatory processes, which means that the dynamic shifts including possible delinking between economic growth and environmental degradation are seldom covered by systematically organized statistics. For instance, a historical Environmental Kuznets Curve (EKC), where cross-sectional data suggest a turnaround income level, corresponding to the leading countries' incomes in, say, 1970 implies that the historical process is not covered by the official metrics. It is therefore important to reconstruct data series for periods for which official statistics do not exist. Historical reconstruction of economic data is in turn a field of research in economic history known as Historical National Accounting. The historical GDP reconstructions collected by Maddison (1991) are even today one of the most commonly used sources for long-term growth studies. One of the first Historical National Accounts were done in Sweden in the 1930s, as recognition of the need to understand economic growth and aggregated demand (Lindahl et al., 1937). This means that the first Historical National Accounts were elaborated well after researchers and practitioners had identified growth as an important economic phenomenon, but before the first official National Accounts had been produced in the 1950s. Again, it would take another 20 odd years before the

Handbook of Green Economics. https://doi.org/10.1016/B978-0-12-816635-2.00001-8

official statistics had produced long-enough GDP series for any serious analysis of long-term growth.

A similar process can be recognized regarding sustainable economic growth. Since the early 1960s, several accounting principles, indices, and metrics have been suggested for measuring what today had been defined as sustainable welfare, taking into account degradation of environmental quality and depletion of natural resources. This includes measures such as Herman Daly's Index of Sustainable Economic Welfare, with a point of departure in the thinking of ecological perspectives first introduced in the 1960s by such economists as Nicholas Georgescu-Roegen and Kenneth Boulding, but also the World Bank's Genuine Savings (GS), grounded in more neoclassical environmental economics by economists such as Partha Dasgupta. Since the 1990s, there have also been guidelines for green national accounting as an extension of the System of National Accounts. Recently, Eurostat has also adapted a system accounting for as economic activities to prevent environmental degradation and to conserve and maintain the stock of natural resources, hence safeguarding against depletion.

The historical environmental accounts presented here include estimates of the Environmental Goods and Services Sector (EGSS), damage cost approaches similar to the concepts used in the System of Integrated Environmental and Economic Accounts (SEEA), and finally, GS.

From the silent spring to environmentally adjusted welfare measures

Following the pioneering works of Rachel Carson's *Silent Spring,* which dealt with the dangers of biocides, Kenneth Boulding published his famous essay "The Economics of the Coming Spaceship Earth," in which he argued that the economic system necessarily needed to adapt to the ecological system's limited pools of resources (Carson, 1966; Boulding, 1966). Thus, Boulding highlighted a positive relationship between economic growth and environmental degradation. Boulding was followed by others, not the least Nicholas Georgescu-Roegen who took the entropy law, a cornerstone of modern physics, as the point of departure for criticism of the standard economic theory, proposing a natural scientific theoretical foundation of the economy as open toward nature (Georgescu-Roegen, 1971, 1977). This is still the basis of ecological economics. The *Limits to Growth* report, published in 1972, put forth empirical evidence for the same basic idea, forecasting the consequences of exponential economic growth in a world with a finite supply of raw materials and a biosphere with a limited capability for assimilating pollutants (Meadows et al 1972). In short, these ecological economists suggested that economic growth was merely an illusion of progress since growth undermined further welfare achievements due to environmental degradation and resource depletion. Several neoclassical economists, arguing that substitution and technological progress would counterbalance resource scarcity, opposed this view (Maurice and Smithson, 1984). Given well-defined property rights, price signals would promote the necessary technology. Other economists, such as Erik Dahmén, pointed at "environmental taxes," an idea originally promoted by Arthur Cecil Pigou in the 1920s, as a mean to put a price on the environment (Dahmén, 1968). While the ecological economists pointed at a negative, linear relationship between growth and a good environment, the neoclassical economists rather suggested a positive one. A third position was suggested in the late 1980s and early 1990s with, first, the WCED report, in which the concept of sustainable development rejected an unavoidable trade-off between growth and the environment, conditional on environmental regulation, social progress, and generally sound institutions (Brundtland et al., 1987). Based on empirical generalizations, the so-called EKC hypothesis suggested a dynamic relationship between growth and the environment, where an initial negative relationship transforms into a positive one, due to the income elasticities of environmental demand and technological change.

Apart from environmental issues, the early 1970s also saw a discussion on the shortcomings of GDP as an appropriate measure of welfare. William Nordhaus and James Tobin were the first ones to suggest and estimate a welfare-adjusted macro aggregate in their Measure of Economic Welfare (MEW), which considered the value of leisure time and the amount of unpaid work, government final expenditures into intermediates, consumption and net investment, and a reclassification of some household expenditures and to some extent environmental degradation (Nordhaus and Tobin, 1972a,b). The latter was estimated as the "disamenity premium," the difference between urban and rural incomes, assumed to reflect a compensation for the disadvantage of pollution in the city. The MEW served as inspiration for further developments of welfare measures. In 1981, Xenophone Zolotas estimated an Index of the Economic Aspects of Welfare (EAW), in which he deducted half of the control costs for air and water pollution, all control costs for solid waste, and all estimated damage costs for pollution (Zolotas, 1981). Thus, the EAW was more comprehensive than the MEW from an environmental point of view, since the EAW included, at least indirectly, depreciation of environmental capital.

A next attempt to measure sustainable welfare was Herman Daly's and John Cobb's Index of Sustainable Economic Welfare (ISEW), which continued in the vein of previous measures, but with a more elaborated theoretical framework (Daly and Cobb, 1990). The ISEW was the empirical measurement of the Hicksian income of the nation, corresponding to the maximum level of consumption, which could be sustained over time. To arrive at a proper consumption measure, the so-called "defensive expenditures," for avoiding declining decreasing welfare, were deducted from traditional consumption. This included, for instance, the cost of commuting. As the Hicksian income assumes an intact total capital, the ISEW also required a proper measure of depreciation of man-made and environmental capital. It was therefore necessary to extend the traditional asset boundary with natural and human capital, in practice attempting to measure net investments in the total capital stock. The ISEW and a subsequent development known as the Genuine Progress Indicator were eventually measured for several countries (see, for instance, Jackson and Stymne, 1996).

There were also similar but less well-known attempts to arrive at welfare and sustainability measures in the 1990s. Peter Rørmose-Jensen and Elisabeth Møllgaard estimated a welfare index (WI) for Denmark in which environmental costs included emissions of SO_2, NO_X, agricultural pollution, lead, and effluents (Rørmose-Jensen and Møllgaard, 1995). Basically, the environmental costs were estimated as a mix of actual and potential avoidance costs, reflecting the reoccurring problem of finding appropriate costs to back up the studies. Outside the domain of social accounting is Arne Jernelöv's measure called the *environmental debt*, which he calculated for Sweden for 1992 (Jernelöv, 1992). The environmental debt was defined as the costs for restoring previous environmental damage to an acceptable level, provided that the damage was reparable. In economic terms, it can be seen as the replacement cost for the part of the damaged environmental capital, which is possible to recreate. Jernelöv did not deal with resource depletion nor any welfare aspects of development.

Integrated environmental and economic accounting

In 1993, a proposal for SEEA was presented as a satellite system to the System of National Accounts (SNA-93). The environmental accounting system shows how different sectors interact with the environment through flows of environmental goods and emissions of residual waste. The basic design of SEEA as a satellite accounting system suggests that it was designed not to disrupt or alter the information within the ordinary SNA. It also makes the SNA suitable as a point of departure for the organization of reconstructed historical environmental data.[1] The main accounting idea is that the environment should be regarded as capital. Environmental damage is thus an analogue to capital depreciation. The SEEA is therefore basically an extension of the asset boundaries used in the SNA. The environmental assets are therefore not included in the traditional national wealth or stock of fixed assets.

The accounting system also shows flows of environmental goods to the economy and flows of residuals (pollutants and similar products like garbage) to the environment. The system elaborates both physical and monetary accounts. Capital is considered as an asset to produce a stream of capital services, which are known as environmental services within the context of the SEEA. In the monetary versions of the SEEA, a new aggregate, environmentally adjusted net domestic product (EDP), is estimated. EDP is calculated by subtracting the costs of natural resource depletion and environmental degradation from net domestic product (NDP). Contributions to NDP and EDP by production sectors are termed value added (VA) and environmentally adjusted value added (EVA), respectively. Estimations of EDP

[1] Methodological discussions in the chapter are based on Lindmark, M. (1998) *"Towards Environmental Historical National Accounts for Sweden 1800–1990. Methodological Considerations and estimates for the 19th and 20th centuries"*, (Diss) Umeå Studies in economic History 21, Umeå.

require the use of imputed environmental costs. This is because the SEEA environmental assets are, per definition, not controlled by institutional units, such as the government, corporations, or household, which implies that there are no environmental asset markets on which asset prices may be formed. The costs occurring due to depreciation of environmental assets may, however, be estimated as either contingent costs, avoidance costs, maintenance costs, or damage costs based on market prices.

Historical environmental accounting based on SEEA approaches

For historical estimates, several valuation approaches may be considered, bearing in mind that limitations on available data and even conceptual issues affect the array of methods that can be used. For instance, contingent costs typically reflect the contemporary valuation of a natural asset, say a population of wild wolves. This means that the valuation is bound in a specific historical context, depending on several factors. Contemporary willingness-to-pay investigations indicate a positive value on wolves. However, this valuation cannot immediately be adopted for pricing of the historical asset value of wolves without paying attention to the institutional context. In Sweden, for example, the government paid bounties on killed wolves up to 1965, implying that wolves were regarded as a nuisance rather than an asset. The asset value was in fact negative until 1966. How the value developed from the mid-1960s until the first contemporary willingness-to-pay investigation is not known, since historical willingness-to-pay investigations are not possible. The problem as such shows how prices are always bound in the specific valuation system of a certain point in time. This is not unique to environmental valuation. Also when economic growth is measured as GDP, we need to anchor the GDP index in a specific base year. Over longer periods, this poses a series of problems also in traditional National Accounting.

Maintenance costs are the costs that would have occurred if a certain environmental quality had been kept in its original status. Maintenance activities could include, for instance, reduction of environmentally damaging economic activities and substitution of inputs for economic activities. Other types of maintenance activities include restoration of the environment and actions taken to lessen the impact of the economic activity. Valuation can either be done on basis of actual or hypothetical cost data. Actual costs are the costs that are associated with keeping an environmental asset intact and could include, for example, undertaken pollution abatement costs. From this follows that the actual avoidance or maintenance costs are, per definition, sufficient. If the environmental asset is indeed kept intact, the environmental costs must already have been internalized in the economy. What the environmental account is showing is then the expenditure needed for sustaining the environmental assets. Hypothetical costs are the costs that would have occurred if the asset was to be kept intact. Here, the underlying assumption is that actual avoidance measures taken are not sufficient to offset the degradation of the environmental capital. This is also the guiding principle behind contemporary accounting for the Environmental Goods and Service sector. Several important economic-environmental processes cannot be studied through contemporary environmental accounts. For instance, it is likely that the actual maintenance costs rise before the turnaround point of an EKC is reached.

When elaborating historical cost estimates, data could either involve early or late base-year estimates of avoidance costs. In many cases, it would, however, be difficult to find early estimates. Such estimates are also affected by several factors bounded in a historical context. In leading countries, the abatement technology was under- or even undeveloped prior to an EKC turnaround. A likely bias in early cost assessments is therefore overestimates of actual costs. On the other hand, the scientific knowledge of environmental damages was typically also undeveloped, tending to cause underestimates of the true costs, as the abatement targets were overly optimistic set, given contemporary knowledge and preferences. Late cost assessments would reflect the costs occurring using a mature environmental technology, which would tend to underestimate the historical cost.

The most likely type of historical indicator is estimated flows of emitted pollutive substances such as sulfur. Such emission and late cost data make it possible to estimate the costs that would have occurred, if the emission levels had been reduced in such a way that the environmental capital stock was sustained, if modern technology had been available. This is a counterfactual cost estimate expressed in the late cost estimates prices, which implies not only a cost level but also a system of relative prices.

The fact that the resulting series is a counterfactual one is, however, not unique to historical environmental accounting. It actually occurs every time a fixed price series is constructed, and becomes, as previously stated, more obvious given a high rate of technical change over the period. The fact that the difference between output series deflated with early and late base years can be used to explore technical change is even known as the Gerschenkron effect, after the Alexander Gerschenkron's famous article on the industrialization of Bulgaria (Gerschenkron, 1962). In the search for what can be called "environmental Gerschenkron-effects," differences between early and late avoidance cost estimates as indicators of environmental technological development would even be an important research task.

Apart from drawbacks associated with cost estimates, hypothetical maintenance costs do not reflect the true environmental damage costs or the actual depreciation of the environmental asset. Even though it would also be possible to try to compensate for the different biases that occur when using modern estimates as the base year for environmental cost estimates, more advanced estimation methods may complicate the interpretation of the series.

Assuming a universal and globally available technology also implies that the same maintenance costs can be used for cross-country comparisons. The great advantage of the approach is that it is in many cases relatively easy to estimate historical emission flows. In addition, there are already several avoidance cost investigations available.

Depletion costs

The most purist form of historical environmental accounting uses imputed environmental costs at market values. As market values presuppose well-functioning markets, they limit the type of environmental costs that can be accounted for to certain depletion cost and degradation of foremost land assets.

Depletion only concerns nonproduced natural assets, namely wild biota, subsoil assets, and water. Depletion may also regard some foreign assets if environmental goods do not reach the country as imported products. In practice, this refers to the use of common global assets such as fish in international waters. The use of produced natural assets is treated as depletion and is also incorporated in SNA-93. A valuation of stocks and depletion is performed according to methods attributed to the three accounting approaches. The use of land, landscape, etc. includes quality changes due to changes in land practices, land use, and degradation of land due to soil erosion as the most important posts. Soil erosion can be estimated through diminishing land values at market prices.

The possibility to use market prices for historical environmental accounts depends on available data. In some cases, such as natural forests, there may even readily available net prices such as stumpage prices. These are the prices for standing trees before logging and transportation. A conceptual problem is that prices are closely associated with institutional control. This may come into conflict with the definition of depletion costs as being associated with nonproduced natural asset, which are not controlled by an institutional unit. Knowledge of institutional conditions is therefore important when constructing historical environmental accounts. Using Sweden as an example, property rights over naturally grown forests in foremost northern Sweden were defined in a series of lawsuits between approximately 1840 and 1906. Stumpage prices from this period may therefore be regarded as the asset prices of timber that had grown outside institutional control, but that, at the time of harvesting, was controlled by an institutional unit. Harvests of institutionally controlled forests such as timber tracts should not be recorded as depletion. This means that the institutional status of the asset affects environmental costs. As the institutional process tends to be incremental, this poses further challenges in terms of environmental accounting. At other occasions, one can imagine a sudden institutional change when the legal status of the forest is instantly changed due to the passing of new legislation. In the Swedish historical environmental accounts, depletion cost for forest assets was only recorded as long as the total standing timer volume was decreasing (Lindmark, 1998). Thereafter, growing forest was treated as capital formation affecting GDP. Occasionally, estimated crude net prices, which refer to timber prices minus extraction costs, are also readily available in the official statistics.

Degradation costs would, in principle, be traceable through land values. Such data can, however, be expected to be highly difficult to find. Depletion of subsoil assets can, for instance, be estimated as resources rents net of production costs, including a normal rate of return on working capital (Dasgupta, 1989).

As pointed out earlier, several valuation methods have been suggested for environmental accounting purposes. For subsoil assets, the net price method was the primary method used in the Swedish historical environmental accounts, where a net price was first calculated as the gross unit price net of the unit extraction costs and a normal unit profit. The normal profit was calculated as the rate of return in other manufacturing industries applied on the working capital in the mining industry. An advantage with the method is that depletion costs only require flow data and recorded assets. However, in the case of the Swedish historical environmental accounts, depletion costs were also estimated using the user-cost-approach (El Serafy, 1989, 1991). The idea behind the approach is to divide the income stream into two parts, where the first part represents the 'true' income of the mining activity. The second part is the part of the total incomes that has to be reinvested in another economic activity to secure the same income stream when the mineral deposit stock is depleted. This means that an interest rate must be chosen and that the extraction-to-reserve ratio must be known. The reserves should correspond to the economically extractable deposits in the beginning of the accounting period. As the economic conditions are due to both long-term change and short-term fluctuations, this estimation is far from being easy. In Sweden, the first estimates of extractable iron ore reserves were made in the late 19th century, reflecting the relatively limited geological knowledge of time (Nordenström, 1893). With better geological methodologies, the reserves were adjusted upward, which was also true as mining technology improved. Collapsing prices, as in the 1970s, meant on the other hand that all central Swedish iron ore deposits were written off. One way of dealing with the problem of fluctuating extraction-to-reserve ratios is to use contemporary estimates of extractable deposits and use them for the historical data.

Expenditures for environmental protection and resource management

While environmental accounting in the vein of SEEA departs from an extended asset boundary, which to a high degree depends on imputed costs, there are also environmental accounting procedures, which depart from the use side of the economy (Eurostat, 2016, 2017). Here, two rather similar versions may be noticed, the EGSS accounts and the Environmental Protection Expenditure Accounts (EPEA). Thus, environmental activities are recorded on the basis of the purpose of final or intermediate consumption and investments. From the supply side of the economy, the corresponding activities are the EGSS. Accordingly, the EPEA and EGSS type of environmental accounting uses actual economic transactions. Instead of extending the asset boundary, the accounts depend on reclassifications of final use. The advantage as compared with the SEEA approach is that imputed costs are avoided, whereas the drawback is a loss of the conceptual connection to sustainability, as there is no ambition to measure the depreciation of the environmental capital stock in the EPEA. Still, it should be recognized that the basic concepts of expenditures for environmental protection and resource management are very close to the avoidance cost approach in the SEEA. There is, however, no ambition to monitor whether these avoidance measures provided by the EGSS are sufficient for offsetting degradation or depletion.

The EGSS became a new European statistical standard during the 2010s. Basically, the sector is defined as economic activities to prevent, reduce, and eliminate pollution and any other form of environmental degradation (environmental protection) and to conserve and maintain the stock of natural resources, hence safeguarding against depletion (resource management. The purpose of EGSS output is therefore, as pointed out, related—but not equivalent—to the concept of strong sustainability: a nondeclining stock of environmental assets.

Reconstructing historical environmental goods and services sector accounts for Sweden

Similar to how traditional consumption is classified in the National Accounts, environmental protection activities are classified according to their main purpose in a system called the *Classification of Environmental Protection Activities* (CEPA), whereas resource management activities are classified according to the *Classification of Resource Management Activities* (CReMA). Table 1.1 provides an overview of CEPA including comments on the historical data availability from 1970.

The message is that it is possible to arrive at historical reconstructions by piecing together information from different types of sources. From 1969 onward, the government undertook occasional investigations of the environmental investments in mining and manufacturing industries. These investigations were not supported by a formalized accounting framework and can therefore not be considered as fully comprehensive. For instance, the investigation aimed at estimating investments in existing facilities, which means that investments in capacity-enhancing projects, which often used superior environmental technology, were only estimated indirectly. This implies that the investigations are biased toward end-of-pipe type of solutions and are to be seen as lower-bound estimates of the true environmental investments. Still, these investigations are the only aggregated information available on early environmental protection activities by the Swedish business life. The early investigations do not separate between air and water protection activities, which means that these activities probably must be aggregated in the historical environmental accounts.

Table 1.1 Classification of environmental protection activities and historical data availability.

Code	Activity	Description	Historical data availability from 1970
CEPA 1	Protection of ambient air and climate	Activities aimed at the reduction of emissions into the air including control of emissions of greenhouse gases and ozone-depleting gases	Embedded in manufacturing industry environmental investments. Certain government programs. Carbon, sulfur, and NO_x tax?
CEPA 2	Wastewater management	Prevention of pollution of surface water. Includes the collection and treatment of wastewater including monitoring and regulation activities Water	National Accounts NA SNR 4410 Will include water works. Will not include activities outside SNR 4410, including manufacturing industry environmental investment
CEPA 3	Waste management	Collection and treatment of waste, including monitoring and regulation activities. It also includes recycling and composting, the collection and treatment of low-level radioactive waste, street cleaning, and the collection of public litter	National Accounts SNR 9200 Sanitary and similar services, except sewage disposal
CEPA 4	Protection and remediation of soil, groundwater, and surface water	Measures and activities aimed at the prevention of pollutant infiltration, cleaning up of soils and water bodies.	Investments in mining dams: mining company records. Activities against acidification of lakes and forests from 1977. Post in the government budget
CEPA 5	Noise and vibration abatement	Activities aimed at the control, reduction, and abatement of industrial and transport noise and vibration	Partly embedded in manufacturing industry environmental investments
CEPA 6	Protection of biodiversity and landscape	Activities aimed at the protection and rehabilitation of fauna and flora species, ecosystems, and habitats	Post in the Government budget
CEPA 7	Protection against radiation	Activities and measures aimed at the reduction or elimination of the negative	Budget for the Swedish Radiation Safety Authority and until 2008 the

Table 1.1 Classification of environmental protection activities and historical data availability.—*cont'd*

Code	Activity	Description	Historical data availability from 1970
		consequences of radiation emitted from any source. Included is the handling, transportation, and treatment of high-level radioactive waste	Swedish Nuclear Power inspectorate. Activity of the Swedish Nuclear Fuel and Waste Management Company, payments from the Nuclear Waste Fund (established in 1982)
CEPA 8	Research and development (R&D)	R&D activities and expenditure oriented toward environmental protection	Government supported R&D as stated in the Government budget

Note: SNR is Statistics Sweden's classification of industrial activities.

CEPA 2, "Wastewater management," can be approximated by using national accounting data for SNR 4410 (water works and supply, including sewage disposal).[2] Wastewater management that was part of industrial processes is not included in the sector but would be embedded in the environmental investments and running expenditures of the mining and manufacturing industries. Furthermore, "wastewater management" underwent significant improvement with regard to the technologies used already from the 1970s onward. The additional cost of improved technology is, however, embedded in the aggregated data for the sector and does not therefore constitute a major accounting problem. SNR 4410 does, however, also include water supply and not only sewage treatment and disposal, SNI 90001, for which data have not been collected separately.

CEPA 3, "Waste management," is also captured in the national accounting data for the sector SNR 9200. Again, it is not directly possible to quantify quality improvements. Waste disposal has, for instance, evolved from a very high share of deposition on garbage dumps in the early 1970s to a high degree of recycling during the 2000s. Waste management services are usually provided by firms within the sector and more seldom as activities within other sectors. SNR 9200 can therefore be expected to capture most water management activities.

Data for CEPA 4, "Measures and activities aimed at the prevention of pollutant infiltration, cleaning up of soils and water bodies," are not recorded separately in the available statistics. Instead, indirect sources of information are necessary. Probably, one of the largest forms of activities in Sweden includes the construction of dams to prevent leakage from mines. These investments can be collected from the leading mining companies, Boldien and LKAB, but are not estimated for this chapter. A

[2] SNR is Statistics Sweden's classification of Industries.

second major undertaking was the program to counteract acidification of lakes. This was a government program in 1977, for which data are included in the government budget.

CEPA 5, "Noise and vibration abatement," is another category for which no data have been collected historically. Some investment may be included in the general environmental investments in mining and manufacturing industries. There are no attempts to reconstruct CEPA 5 in this chapter.

CEPA 6, "Protection of biodiversity and landscape," has been a specific post in the government budget since the 1970s. This includes expenditure for the acquisition and management of ecologically valuable land and biotopes. Costs for protection of biodiversity and landscape carried by, for instance, forest companies and NGOs have not been estimated for this chapter.

CEPA 7, "Protection against radiation," has, in Sweden, mainly been the responsibility of the Swedish Radiation Safety Authority, whose activities have mainly focused on regulation, surveillance, and activities related to the prevention of technological hazards in nuclear plants. Of these, prevention of technological hazards should not be included in CEPA 7. Still, as it is impossible to separate this activity from the ones supposed to be part of CEMA 7, we have include all the activities of the Swedish Radiation Safety Authority in CEPA 7. Actual handling, transportation, and treatment of high level radioactive waste have been undertaken by SKB, the Swedish Nuclear Fuel and Waste Management Company, which should be fully included in CEPA 7. Another source of information is the Nuclear Waste Fund, which has financed activities related to protection against radiation since 1982.

To start with, CEPA 8, "Research and development," may be embedded in traditional government-funded research and business financed R&D, which makes it close to impossible to account for the true expenditures. In this chapter, we therefore only estimate expenditure on R&D activities, which were financed through specific government-funded research programs on environmental issues. The CEPA 8 is therefore a lower bound estimate. Table 1.2 shows a corresponding overview of the Classification of Resource Management Activities (CREMA), along with brief comments on how historical reconstructions may be performed.

Management of water, which focuses on various measures to reduce inland water intake, is not relevant in the Swedish context, since the issue does not pose a major problem due to hydrological conditions. CREMA 11a, dealing with management of forest areas is, however, of major importance, and the official forest statistics include comprehensive statistics on expenditures for replenishment activities. Minimization of the intake of forest resources (CREMA 11b) is more difficult, since it deals with efficiency and productivity issues that are, normally, part of the ongoing innovation process in the forest industry (Söderholm and Bergquist, 2013). For this reason, estimates of CREMA 11b have not been undertaken. Activities aimed at the minimization of the intake of wild flora and fauna have not been a substantial activity. Due the limited size of the sector, we have not estimated data.

Heat/energy saving and management in CREMA 12 mainly focuses on energy recycling from garbage and waste. Here, the official data are comprehensive from 1990, noticing that the ban on landfills was introduced in 1998. Production of energy

Table 1.2 The CREMA classification of resource management activities.

Code	Activity	Description	Sources
CREMA 10	Management of water. Minimization of inland waters intake	Activities aimed at the minimization of inland waters intake	Not relevant in a Swedish context
CREMA 11a	Management of forest areas	Replenishment activities or development of new forest areas, raining, and information and general administration activities. Activities concerning the protection and restoration of forests as habitats, ecosystems, and landscapes are excluded	Replenishment activities are recorded in the official forest statics and are used for estimates of the activity
CREMA 11b	Minimization of the intake of forest resources	Reduction of the intake through in-process modifications related to the reduction of the input of timber resources for the production process	It is part of the ongoing activities in the industry and is not accounted for separately
CREMA 12	Management of wild flora and fauna	Activities aimed at the minimization of the intake of wild flora and fauna	Included under CEPA 6
CREMA 13a	Production of energy from renewable sources	Renewable (nonfossil) energy includes wind, solar, aerothermal, geothermal, hydrothermal and ocean energy, hydropower, biomass, landfill gas, sewage treatment plant gas, and biogases	Energy statistics in combination market prices are used to estimate the activity
CREMA 13b	Heat/energy saving and management	energy recovery from nonrenewable sources (e.g., nonbiodegradable waste)	Data on energy production from waste. Some government programs aiming at energy saving
CREMA 13c	Minimization of the intake of fossil energy resources for other uses	Activities aiming at the minimization of the intake of fossil energy resources for uses other than energy production	No historical estimates are done for CREMA 13

Continued

Table 1.2 The CREMA classification of resource management activities.—*cont'd*

Code	Activity	Description	Sources
CREMA 14	Management of minerals	Activities aiming at reducing the use of mineral resources. Recycling of scrap metals	It is part of the ongoing activities in the industry and is not accounted for separately, apart from Recycling of scrap metals
CREMA 15	Research and development activities for natural resource management	R&D for renewable energy, for energy and minerals savings, for timber and other biological resources savings and so forth	Some government programs aiming at energy saving

Source: EUROSTAT, 2016. Environmental Goods and Service Sector Accounts. Practical Guide. 2016 Edition. Luxembourg: Publications Office of the European Union; EUROSTAT, 2017. Environmental Protection Expenditure Accounts. 2017 Edition. Luxembourg: Publications Office of the European Union.

from renewable sources (CREMA 13a) includes hydropower, wind power, and biofuels. The estimates for the sector depart from energy statistics and price data. Heat/energy saving and management in CREMA 13b mainly focuses on energy recycling from garbage and waste for which data are available from 1990. CREMA 13c focuses on activities aiming at the minimization of the intake of fossil energy resources for uses other than energy production, which is a small sector in Sweden, and is therefore not estimated.

Results of ongoing estimates

Estimating both SEEA and EGSS provides analytical tools for examining basic environmental and ecological relationships. Fig. 1.1 shows carbon emissions estimated using three different approaches (Lindmark and Acar, 2013). First, the emissions, expressed in physical units, have been monetarized, using Nordhaus' conservative estimate at 30 USD per ton in the year 2000. Second, the price has been discounted, allowing it to grow at the discount rate used by Nordhaus. Finally, the carbon series have been transformed to a current price series, by reflating the discounted price with the investment deflator from the National Accounts.

As expected, the series show distinctly different growth patterns. The fixed price series, using the year 2000 price reference, reflects the actual carbon emissions, characterized by a first decrease from the early 1970s until 1990 and a second period of falling emissions from 2003 to 2016. The discounted volume series is relatively constant, implying that the carbon emissions, measured in physical units, have decreased at approximately the interest rate assumed by Nordhaus, which is lower than the Swedish GDP growth rate.

Fig. 1.2 shows the damage cost of other pollutants, along with the environmental protection expenditures, both expressed in the prices of the year 2000.

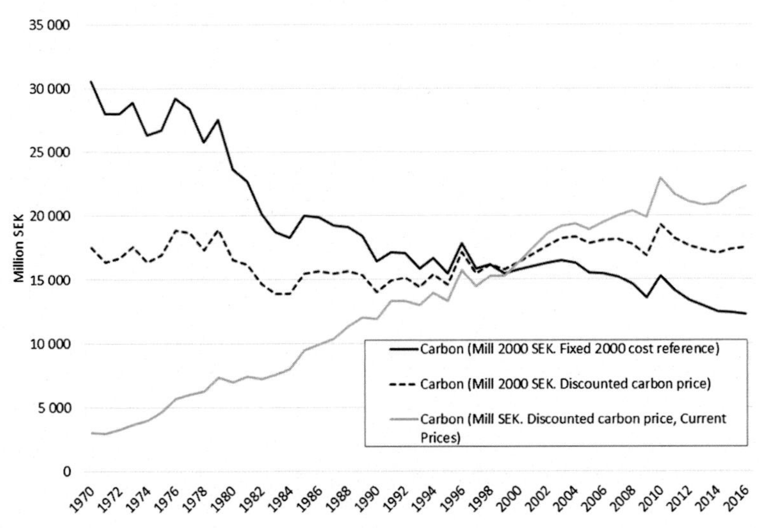

FIGURE 1.1

Carbon damage cost in constant 2000 prices, constant discounted 2000 prices, and discounted current prices. Sweden 1970–2016.

Source: Own calculations based on Lindmark, M., Acar, S., 2013. "Sustainability in the making? A historical estimate of Swedish sustainable and unsustainable development 1850–2000". Ecological Economics 86, 176–187.

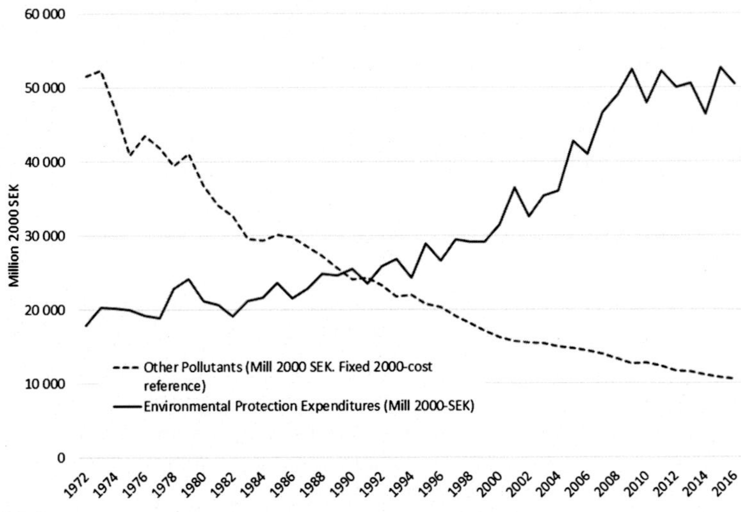

FIGURE 1.2

Pollution damage costs and environmental protection expenditures. Constant 2000 prices. Sweden 1970–2016.

Source: Lindmark, M., Acar, S., 2013. Sustainability in the making? A historical estimate of Swedish sustainable and unsustainable development 1850–2000. Ecological Economics 86 176–187, own calculations as described in the text.

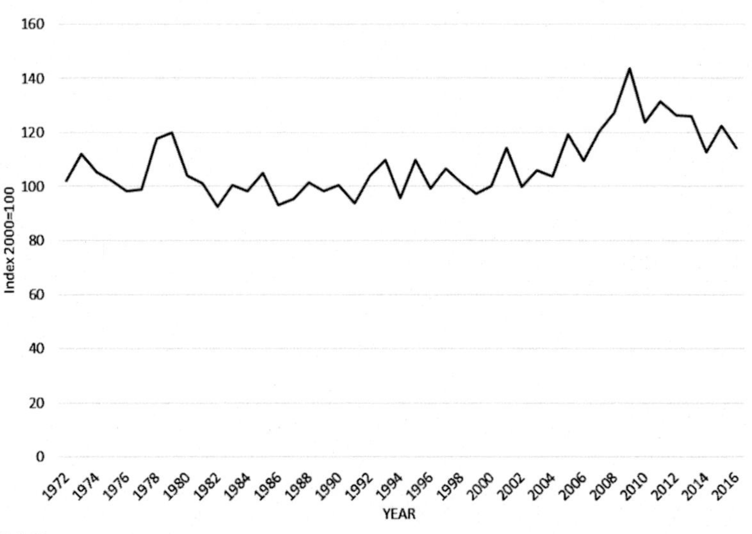

FIGURE 1.3

Environmental protection expenditure ratio to GDP. Sweden 1972–2016. Index 2000 = 100.

Source: Own calculation.

As expected, pollution damage costs decreased over the period, whereas environmental protection expenditures have increased until a plateau was reached in 2009. At the same time, environmental protection expenditures constituted a relatively constant share of GDP between 1972 and 2006. Thereafter the environmental expenditure share seems to have jumped up to a new level, approximately 20% higher than the previous one. The relatively constant cost share is suggesting Cobb-Douglass characteristics, implying constant marginal utility and constant returns to scale may apply to environmental consumption (Fig. 1.3).

Conclusion

Although historical environmental accounting presents several methodological challenges, it still provides insights on the behavior of environmental expenditures and damage costs on an aggregated level. In this chapter, it was demonstrated how different environmental accounting approaches can supplement each other to provide an overall interpretation of the interplay between economy and the environment. To start with, the social costs of traditional pollutive emissions have declined since the early 1970s. Secondly, the expenditure on various economic activities intended to reduce environmental degradation and depletion costs has increased over the same period. Actually, the expenditures may have increased at more or less the same rate as GDP, implying support for the argument that economic growth indeed creates more resources for environmental protective expenditure.

The constant cost share also suggests that environmental expenditures may reveal Cobb-Douglass properties. If so, the marginal utility of environmental expenditures may be constant. This means that it may be difficult to mobilize broad, public support for radically more ambitious environmental policies. From this follows that policymakers may need to put more emphasis on priorities between different environmental goals. Another interesting result is that carbon emissions have fallen, but that the social cost in discounted, current prices has remained more or less constant. This means that Swedish carbon emissions have fallen at the same rate as the interest rate used by Nordhaus for the present value calculation of carbon. This may of course be a coincident, but it could also be an example of a transition of carbon emission along a socially and environmentally stable path, an issue that would require additional research.

References

Boulding, K.E., 1966. The economics of the coming spaceship earth. In: Jarrett, H. (Ed.), Environmental Quality in a Growing Economy. Resources for the Future/Johns Hopkins University Press, Baltimore, MD, pp. 3–14.

Brundtland, et al., 1987. Report of the World Commission on Environment and Development: Our Common Future. Oxford University Press, Oxford.

Carson, R., 1966. Silent Spring. Hamilton, London.

Dahmén, E., 1968. Sätt Pris På Miljön: Samhällsekonomiska argument i miljöpolitiken. SNS-förlag, Stockholm.

Daly, H.E., Cobb, J.B., 1990. For the Common Good. Green Print, London.

Dasgupta, P., 1989. Exhaustible resources. In: Friday, L., Laskey, R. (Eds.), The Fragile Environment. New Approaches to Global Problems. Cambridge Univ. Press, Cambridge, pp. 107–126.

El Serafy, S., 1989. The proper calculation of income from depletable natural resources. In: Ahmad, Y.J., E1 Serafy, S., Lutz, E. (Eds.), Environmental Accounting for Sustainable Development. The World Bank, Washington D.C., pp. 10–18

El Serafy, S., 1991. The environment as capital. In: Costanza, R. (Ed.), Ecological Economics: The Science and Management of Sustainability. Columbia University Press, New York, pp. 168–175.

EUROSTAT, 2016. Environmental Goods and Service Sector Accounts. Practical Guide, 2016 Edition. Publications Office of the European Union, Luxembourg.

EUROSTAT, 2017. Environmental Protection Expenditure Accounts, 2017 Edition. Publications Office of the European Union, Luxembourg.

Georgescu-Roegen, N., 1971. The Entropy Law and the Economic Process. Harvard University Press, Cambridge, Massachusets.

Georgescu-Roegen, N., 1977. The steady state and ecological salvation: a thermodynamic analysis. BioScience 27 (4), 266–270.

Gerschenkron, A., 1962. Some aspects of industrialization in Bulgaria, 1878–1939. In: Gerschenkron, A. (Ed.), Economic Backwardness in Historical Perspective. A Book of Essays. Belknap Press, Cambridge, Massachusetts, pp. 198–234.

Jackson, T., Stymne, S., 1996. Sustainable Economic Welfare in Sweden. A Pilot Index 1950–1990. Stockholm Environment Institute, Stockholm.

Jernelöv, A., 1992. Miljöskulden. En rapport om hur mijöskulden utvecklas om vi ingenting gör. SOU 1992:58. Allmänna förlaget, Stockholm.

Lindahl, E., Dahlgren, E., Kock, K., 1937. National Income of Sweden 1861–1930. P.S. King & Son, London.

Lindmark, M., 1998. "Towards Environmental Historical National Accounts for Sweden 1800–1990. Methodological Considerations and Estimates for the 19th and 20th centuries",(Diss) Umeå Studies in Economic History 21. Umeå University, Umeå.

Lindmark, M., Acar, S., 2013. Sustainability in the making? A historical estimate of Swedish sustainable and unsustainable development 1850–2000. Ecological Economics 86, 176–187.

Maddison, A., 1991. Dynamic Forces in Capitalist Development. A Long-Run Comparative View. Oxford University Press.

Maurice, C., Smithson, C.W., 1984. The Doomsday Myth. 10 000 Years of Economic Crises. Hoover Institution Press.

Meadows, D.H., Meadows, D.L., Randers, J., Behrens III, W.W., 1972. Limits to Growth. (London).

Nordenström, G., 1893. Sveriges Malmtillgångar. Jernkontorets annaler, pp. 198–211, 1893.

Nordhaus, W.D., Tobin, J., 1972a. Economic Growth and Declining Social Welfare. National Bureau of Economic Research, New York.

Nordhaus, W.D., Tobin, J., 1972b. Is Growth Obsolete? National Bureau of Economic Research, New York, 96.

Rørmose-Jensen, P., Møllgaard, E., 1995. Measurement of a Welfare Indicator in Denmark 1970–1990. Rockwell Foundation Research Unit, Copenhagen.

Söderholm, K., Bergquist, A.K., 2013. Growing green and competitive—a case study of a Swedish pulp mill. Sustainability 5 (5), 1789–1805.

System of National Accounts 1993, 1993. Commission of the European Communities, International Monetary Fund., Organisation for Economic Co-operation and Development., United Nations. World Bank, Brussels, Luxembourg, New York, Paris, Washington D.C.

Zolotas, X., 1981. Economic Growth and Declining Social Welfare. Bank of Greece, Athens.

CHAPTER

How to handle natural capital within the context of the green economy?

2

Cristián Ducoing

Introduction

Climate change is a reality of our time, and the question is not how to hold it back, instead the real question is *how much* we can hold it back. The recent reports from international organizations and national agencies are quite clear: if we do not take severe actions in the next years, the effects of global warming are going to be irreversible.[1]

Our resource-intensive production model is the main reason behind global warming. The enormous amount of pollutants resultant from two centuries of industrial progress have increased exponentially the amount of CO_2 in the atmosphere with serious consequences for the environment. This cost was not acknowledged previously, and the past generations have been using finite resources as an infinite. The low value given to natural capital is one of the explanations of this behavior (Barbier, 2011a, 2011b, 2015; Hanley et al., 2015; Helm, 2015; Hartwick, 1991, 1990).

There are several actions implemented by policymakers to confront climate change and its consequences. The Sustainable Development Goals (SDGs) are an ambitious attempt to achieve 17 objectives by 2030 with the main aim of eradicating extreme poverty without putting the environment at risk for future generations. One of the most striking points from SDGs is related with climate policy. The problem is that global warming is correlated with our current model of economic growth and its intensive use of natural resources. The present measures of development are overwhelmingly dominated by gross domestic product (GDP). It seems that GDP is out of the question or, at least, politicians and policymakers are not making the efforts to replace it in the short term (Coyle, 2015). This GDP "dictatorship" makes the adoption of more sustainable measures through a cleaner and respectful model with our environment difficult (Hanley et al., 2015). The right evaluation and measure of natural capital can be one of the solutions and tools to combine economic development with sustainability. In brief, our economic model (without political sign distinction) has abused the environment by consuming finite resources without any contemplation in intergenerational solidarity.

[1] The report of the Intergovernmental Panel on Climate Change (IPCC) gave several warnings on the difficulties to achieve the Paris Agreement regarding the 1.5°goal (Climate Change, 2018).

Handbook of Green Economics. https://doi.org/10.1016/B978-0-12-816635-2.00002-X

The aim of this chapter is to contribute to the debate giving some insights from economic and environmental history in the difficult (but unavoidable) challenge to combine economic progress and sustainability. If we have to combine both objectives, a rational choice is to give to natural capital a fair and economically viable valuation. The chapter is organized as follows: The Previous research section presents a brief literature review on natural capital measurement. Current measures including natural capital section deals with the current measures related with natural capital, such as natural wealth, green accounting, and genuine savings (GS). Natural capital as base of the green economy section proposes solutions to include natural capital in standard and heterogeneous measures of development. Conclusions and policy lessons section concludes with some policy lessons that could be implemented.

Previous research

The valuation of natural capital is part of the history of national accounting but with great differences regarding the timing, the concept itself, and the importance given to the topic. Since the 19th century, extremely related to the Industrial Revolution and the change from organic to fossil fuels, businessmen, politicians, economists, and policymakers have tried to value natural capital understood as *natural resources*. Jevons's concern about the coal reserves in Britain and the spectacular rate of depletion that this resource had undergone during the years of Industrial Revolution is well known (Jevons, 1866). In the same line, several authors of the 19th century were concerned about the negative effects brought by smoke and pollutants (smog) from fabrics in the health of urban citizens. However, if we delimited the concept of natural capital to traditional and standard national accounting, the value of resources and biodiversity changes completely. Natural resources and biodiversity were taken as granted and never had a special consideration until the 1950s. Some changes that occurred in this decade can be linked to the Cold War and the needs for resources in a different geopolitical context. The creation of *Resources for the Future* commission is a landmark change regarding natural resources and environmental concern by policymakers and governmental institutions.[2] One of the results of this organization was the book *Scarcity and Growth. The Economics of Natural Resource Availability* by Barnett and Morse (1963). The aim of the book is directed specifically to understanding the effects of natural resources scarcity and its influence in the socioeconomic context.[3] As we can see, natural capital as a part of the

[2] "Resources for the Future is a nonprofit corporation for research and education in the development, conservation, and use of natural resources" (Barnett and Morse, 1963, title page).

[3] Our main effort, therefore, is directed to a thorough examination of the conceptual and empirical foundations of the doctrine of increasing natural resource scarcity and its effects (Barnett and Morse, 1963, p. 3).

economy became to be considered in the early years after World War II. However, these considerations did not enter the mainstream models and public policies because the GDP predominance did not leave space to "capital," and less to environmental aspects.

The concerns about the effects of economic growth on natural capital became noticeable in the 1970s. Several publications (Peskin, 1972, 1976) and international meetings such as the Club of Rome and the "United Nations Conference on the Human Environment" in Stockholm were alert voices on the dangerous trend of environmental depredation.[4] All these voices had a greater impact with the oil crisis of 1973 and its effects on developed economies. The quick return of low oil prices reduced the concern on limits to growth, but the theoretical field had been created. Besides these issues, a new framework on natural resource depredation and environmental concerns begun in the second half of the 1970s. Works such as Hartwick (1977) gave new insights into the problems derived by resource extraction and environmental externalities from economic growth. From this work derived the so-called *Hartwick's rule*, which defines the amount of physical capital needed to replace the use and exploitation of natural resources.[5]

In the past 25 years, the debate on natural capital has been linked to the critics of GDP as measure of welfare (Barbier, 2015; Stiglitz et al., 2009; Hartwick, 1991). GDP was conceived as *whole* measure of output based on quantity instead of quality (Coyle, 2015). The amount of iron expressed in the statistics did not take into account if it was scrapped, pure, second hand, and so on, the most important characteristics are quantity and *value*. The forest included in the measures of GDP has a similar problem. If we count the output of forest in chips or tons, we are overlooking the main feature of forests: its ecological services. This problematic with National Accounts has been present during the greater part of the 20th century, and it has promoted an economic system with minimum respect to the environment.

The alternatives to GDP that have included natural capital are genuine savings (GS) and adjusted net savings (ANS). Over the past quarter century, GS has emerged as an important indicator of sustainable development. It is based on the concept of wealth accounting, and it is argued that it addresses shortcomings in conventional metrics of economic development by incorporating broader measures of saving and investment (Stiglitz et al., 2009) GS measures year-on-year changes in total capital stocks (physical, natural, social, institutional, and human). Following the studies of Pearce and Atkinson (1993) and Hamilton (1994), the World Bank has published estimates of GS from the mid-1990s to the present. Hamilton and Clemens (1999) and World Bank (2006, 2011, 2018) illustrate the nature of these estimates for almost all countries in the world and show how a negative GS indicator can be interpreted as a signal of unsustainable development in the medium range. In the next

---

[4] Both meetings were held in 1972. The main publication from the Club of Rome meeting in 1972 was reedited in 2012. *Limits to Growth, the 30 year update* (Meadows and Randers, 2012).

[5] A detailed description of this rule is given in Withagen and Asheim (1998).

section, a detailed description of the implications of GS measures for further development of natural capital accounting is discussed.

Current measures including natural capital

Designed to quantify the monetary value of all goods and services entering into market exchange, GDP is often considered the "invention of the 20th century" (Coyle, 2015). However, as we have seen in this book's introduction and in previous sections, there is a growing understanding that maximizing year on year growth in GDP is not a feasible target for the 21st century where "sustainable development" is the key to global survival (Stiglitz et al., 2009).

GDP flaws are quite acknowledged by the new welfare theory and environmental economics. However, GDP measures are so incorporated in public debates and government policies that alternative measures have a long way to go to generate a consensus on the indicators after GDP. Some sporadic and isolated initiatives have been trying to contemplate alternatives or environmentally friendly complements to GDP. The most used and standardized measures could be summarized in natural wealth with green accounting and GS, the latter being the extension of the former.

Natural wealth and green accounting

The concept of natural wealth has been broadening in the past 30 years. Natural wealth is intrinsically linked to the stock of natural resources that a country or region has in a determined period (mainly years). As Helm (2017) rightly pointed out, the role of assets (wealth) has been excluded or overlooked in the economic theory, and this situation is even more notorious with natural capital. As we have mentioned in the introduction and the literature review, the main focus of economic growth is the maximization of flows, GDP being the mainstream indicator for this objective. The System of National Accounts (hereafter, SNA) has been focused on the measure of outputs, giving a secondary role to capital stock, in its several variables (machinery and equipment, infrastructure, housing, etc.). The role of natural capital is one of the great absences in the national accounting measures. The inclusion of natural assets in the SNA has been slow, and a majority of countries do not have a standardized measure of natural wealth in their accounting. A recent effort by the World Bank (WAVES) has pointed out the national initiatives regarding natural capital accounting and how these measures could be included in a global indicator.[6]

How is natural capital measured in our times? It depends on the country and the institutions behind the valuation methodologies. Taking into account the last debates, the price on nature should be estimated with consideration to its economic,

[6] http://www.worldbank.org/en/news/video/2013/04/18/natural-capital-accounting-in-action. In this video, a brief description of the project is given (accessed: 2019/02/01).

ecological, and intergenerational services. In the first place, the price of the economic value of natural capital such as minerals, waters, and forests. These valuations are quite straightforward and are linked to the international commodity market.

Two main conceptions of natural capital valuation-sustainability are defined: strong natural capital rule and weak natural capital rule. The first one assumes that natural assets are *irreplaceable* and its value/volume should be kept at least constant with the investments from natural resources depletion coming back to renewable natural capital. The second rule consists in constant natural capital and rents for depletion distributed in *general* capital compensation. These two definitions are the so-called "strong versus weak sustainability" debate, to be analyzed in the next section.

Genuine savings

A current use of natural capital is its inclusion in the GS indicator. The focus of this kind of measurement is inclusive/comprehensive wealth to account for changes in our capability (measures by different stocks of wealth) to generate future well-being. The central idea is that positive or negative changes in inclusive (comprehensive) wealth are indicators of sustainable or unsustainable economic activity. Both the United Nations and the World Bank have been advocates of more inclusive/comprehensive measures of wealth. The central idea is to focus on changes in wealth rather than on growth in income; Managi and Kumar (2018) highlights the contradiction between changes in wealth and GDP growth rates, whereby GDP growth rates are not mirrored by growth in wealth indicators. However, the focus on inclusive/comprehensive wealth is a relatively recent phenomenon. The United Nations only began publishing inclusive wealth reports from 2012. The World Bank has a longer track record of publishing estimates of ANS or GS, dating from the 1990s, with estimates going back to the 1970s. However, as economic activity existed before these estimates and knowing more about how these concepts map onto the historical record can tell us much about the validity of the concepts themselves as well as about our historical development.

Current estimates of GS contain gross and net saving/investment, depreciation of natural capital, education expenditure, total factor productivity (TFP), and pollutants such as CO_2 emissions. Currently, there are some long-run GS estimates that are worth mentioning. In the case of national examples, the most ambitious is the article by Greasley et al. (2016), covering the GS of the United Kingdom from the 18th century Industrial Revolution to the present. In the same line and methodology, Hanley et al. (2015, 2016) presented estimates for the United States, Germany, and the United Kingdom since 1850. Sweden has long-run estimations, thanks to the work of Lindmark and Acar (2013) and Lindmark (1998). However, the methodology utilized in this chapter differs from the aforementioned papers. Regarding developing countries, del Mar Rubio (2004) has estimated GS for Mexico and Venezuela. The most ambitious estimates for a wider set of countries cover 11 countries throughout the 20th century (Blum et al., 2017).

As we can see in Fig. 2.1, the differences between GDP per capita and GS per capita are extreme (Table 2.1). Before analyzing these two variables and the reasons behind their differences, it is worth mentioning that we are comparing flow measures (GDP per capita) and stock indicators (GS).

Latin American countries are a good example of the effects of natural capital depreciation over development and how economic growth based on natural resource exploitation is a short-run policy. Table 2.2 presents the average participation of several indicators of wealth in the GDP per capita. The Latin American countries in this table are intensive on natural resources. Brazil, Mexico, and Colombia have a noticeable oil and gas industry, Chile is a mining economy with an important share in the copper world market, and Argentina is recognized as an agroindustrial power. The natural resources endowments of these countries should be an important boost for development, but the results in 100 years are extremely disappointing.

Chile, considered as one of the most successful countries in the region, had a poor performance when we take into account the environmental side of the history. In Figure 2.1B and Table 2.2, Chile shows a quite impressive GDP per capita growth in the period 1987–98. If we compare these figures with its sustainability measures, the pictures are completely different. Even including the TFP in the GS indicator, during the greater part of the 20th century, the percentage of savings measured including some kinds of environmental aggregate were negative. In the case of green investment, the percentage is −5.6, an enormous depletion of natural resources if we see net investment as a positive 2.1 during the same period of time. How is this possible? The answer is quite simple: If we account for environmental depredation, weak sustainability is not achieved in a country with a huge exploitation of natural resources. The lessons from these figures are quite relevant to the debate on natural capital. It does not matter which indicator we choose; if we account for natural capital, countries' performances change radically. Any indicator thinking in a green economy should have included natural capital, and the chosen measure must be internationally standardized to allow comparisons.

Natural capital as the basis of the green economy

Any future development in the world economy should focus on the broad concept of sustainability. Probably, the definition of sustainable development from the Brundtland Report is the most appropriate to the challenges that we are facing.

> *Sustainable Development is development that meets the needs of the present without compromising the ability of future generations to meet their own needs. It contains within it two key concepts: the concept of 'needs', in particular the essential needs of the world's poor, to which overriding priority should be given; and the idea of limitations imposed by the state of technology and social organisations on the environment's ability to meet present and future needs.*
>
> **World Commission on Environment and Development (1987, p.43).**

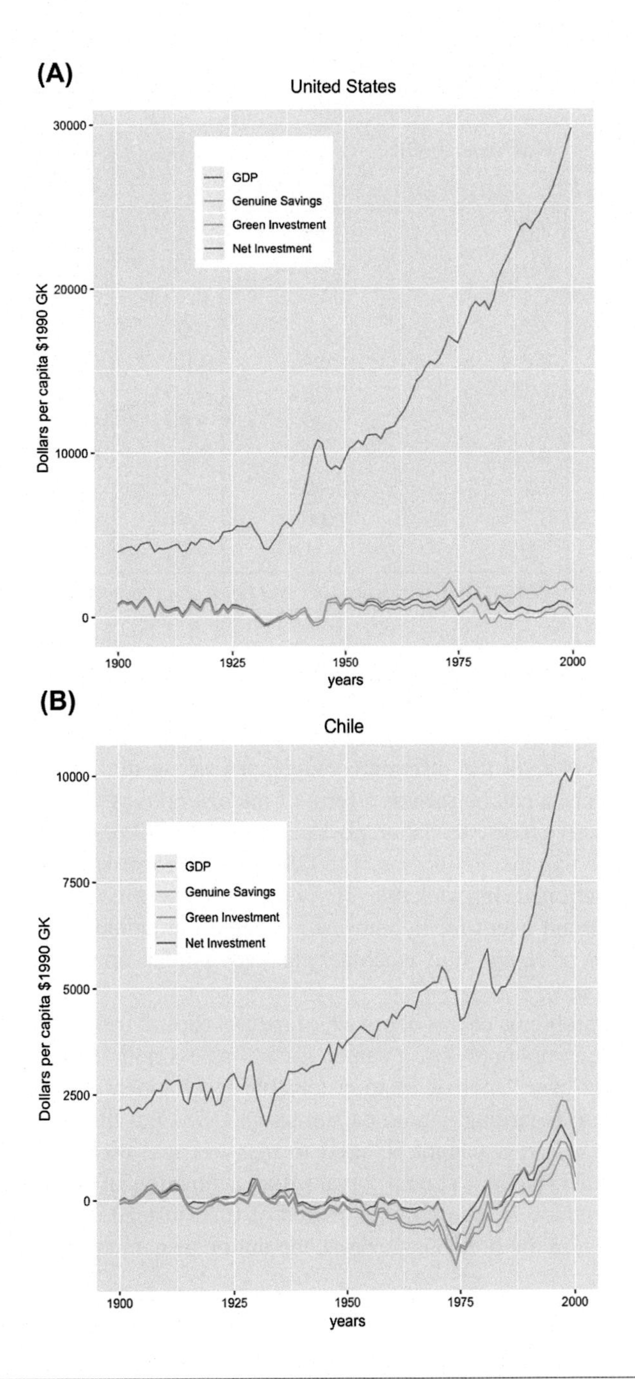

FIGURE 2.1

Genuine savings in comparison with GDP. Chile and the United States, 1900–2000.

Source: Blum, M., Ducoing, C., McLaughlin, E., 2017. National Wealth: What Is Missing, Why it Matters. Oxford University Press, pp. 89.

Table 2.1 Genuine savings growth rates in selected countries.

1900–2000				
	Net investment %	Green %	GS %	GSTFP %
Britain	4.6	2.8	5.5	28.6
Alemannia	9.4	8.1	11.3	49.6
United States	7.1	4.4	8.1	32.6
Australia	7.7	4.3	6.5	24.7
France	8.9	8.4	11.8	29.1
Switzerland	14.1	14.4	17.5	45.4
Argentina	3.7	3.2	3.5	5.1
Brazil	11.1	3.9	5.5	8.4
Chile	2.1	−5.6	−4.1	−2.3
Colombia	8.7	3.6	4.4	4.7
Mexico	8.6	−1.4	−0.9	2.2

GS, *Genuine savings;* GSTFP, *Genuine savings with total factor productivity augmented.*

This definition is under the debate between strong versus weak sustainability that we mentioned above briefly. The weak approach to sustainability links future well-being with changes in wealth (Pearce, 2002). The neoclassical approach to weak sustainability understands the different components of wealth as interchangeable.[7] Future consumption can be seen as a form of interest on past wealth accumulation, since the productive basis, i.e., labor, physical, and intangible capital, are the productive forces used to generate income. The GS approach to sustainability rests firmly on the aforementioned Hartwick rule (Hartwick, 1977), as this shows how consumption can be constant over time by reinvesting rents from natural resource extraction into other forms of capital (i.e., machinery and equipment, infrastructure, or education/human capital).

One of the problems of this approach of natural capital inside GS estimations is that, under certain assumptions, it can be used to assess both the capabilities-based and the outcome-based approaches to sustainable development (Hanley et al., 2015). Another problem regarding natural capital under GS is not firmly grounded in the SNA framework, and it cannot be used to measure and compare countries in a consistent manner. Natural capital is valued as a function of the economic profit that it could produce in a determined period of time instead of its intrinsic value. As we have seen in the previous sections, the notion of natural capital has changed

[7] See Hanley et al. (2015) for a comprehensive review.

Table 2.2 Latin America sustainability indicators mean net investment, green investment, and genuine savings as a percent of GDP.

	NNS	NNSRr	$GSCO_2$	GS	GSTFP	GreenTFP
1900–2000						
Argentina	4%	3%	3%	3%	5%	5%
Brazil	11%	4%	4%	6%	9%	7%
Chile	2%	−6%	−6%	−4%	−2%	−4%
Colombia	9%	4%	3%	4%	5%	4%
Mexico	9%	−1%	−2%	−1%	2%	1%
1946–2000						
Argentina	4%	3%	3%	3%	6%	5%
Brazil	10%	4%	4%	7%	11%	8%
Chile	1%	−9%	−10%	−7%	−5%	−7%
Colombia	15%	9%	9%	11%	11%	9%
Mexico	14%	7%	6%	8%	13%	11%
1970–2000						
Argentina	3%	2%	1%	2%	5%	5%
Brazil	13%	8%	7%	10%	17%	13%
Chile	3%	−8%	−8%	−5%	−3%	−6%
Colombia	18%	11%	11%	13%	14%	11%
Mexico	14%	6%	5%	7%	14%	12%

NNS, *National net savings;* NNSRr, *Net national savings minus resources;* $GSCO_2$, *Genuine savings with CO_2 costs included;* GS, *Genuine savings without CO_2;* GSTFP, *Genuine savings with total factor productivity augmented.*

Source: Blum, M., Ducoing, C., McLaughlin, E., 2017. National Wealth: What Is Missing, Why it Matters. Oxford University Press, pp. 89.

based on its economic utility. The future will require a conception of natural capital as a nonreplaceable asset.

Could natural capital be replaced?

In the modern world, a nation could become rich and developed without natural resources exploitation of their own endowments. Any country can sustain its development through the purchase of commodities (energy and other raw materials) from the external market (Barbier, 2015, p. 55). However, this model has a limitation. Even if the natural resource depletion occurs in a foreign territory, the amount of resources on earth declines and, consequently, the *global biodiversity.* In the green economy framework, natural capital cannot be replaced and *the low cost* that biodiversity has in the international markets is not sustainable any more.

The future needs a green economy, and the recent calls for a green new deal, carbon tax, and new measures of economic development are in the same line of thinking.[8] We cannot continue with a development model based just on growth if we are putting at risk the environment for future generations.

For instance, in tropical regions, there is a steady trend of forest conversion into cropland with its serious and permanent repercussions on global warming. Under the weak sustainability framework, this could be interpreted as positive, because natural capital is constant and nondeclining. However, biodiversity is being threatened, and the environmental capacity is seriously affected. A decoupling between cropland and deforestation in the Amazon regions has been observed in the late 2000s, but it is still early to make conclusions from these observations (Macedo et al., 2012).

Conclusions and policy lessons

The debate on natural capital is open, and a consensus is far away to be reached. Some alternatives, such as GS, have been described in this chapter. This indicator, GS, is a value alternative, but in its current formula it is not the main solution. If natural capital can be replaced for physical capital or other kinds of natural capital without taking into account biodiversity, the prospects for our environment are dark.

Society could benefit greatly from alternative socioeconomic indicators, displacing or complementing the current ones focused mainly on income. Comprehensive measures of development can provide a wider lens to help society escape from myopic logistics. The lower volatility of natural wealth in comparison with income indicators could help to concentrate the discussion on the long run, on the lasting effects of our economic model, and visualize the harmful consequences of some economic activities. Especially for countries centered on natural resources exploitation, natural capital valuation could shed light on the real gap with developed countries. One additional benefit is related to the global measures we are promoting. SDGs require aggregate measures of development, and the country's national level estimations should be expanded. These improvements require more countries and a homogenous methodology as have been mentioned in the introduction. Furthermore, climate action, as example, is better understood when the metrics used to explain the effects of climate change address directly to harmful economic activities. A global natural capital measure shows how the intensive use of natural resources has a dark side, depleting our reserves without improving our quality of life.

To include natural capital in the context of the green economy, some steps should be taken in the medium and short term: first, a standardized method of natural capital accounting allows policymakers to compare and establish future scenarios in economic development plans; second, biodiversity as a global asset should be included

[8] In the case of the green new deal, the extraordinary weather vents of 2018 have meant a noticeable resurgence of this idea (Barbier, 2019).

in future treaties between states—the cost of preserving our natural environment is localized in the countries, but the benefits are enjoyed globally; and finally, a major effort in strong sustainability should be taken.

The lessons are straightforward: current investment in physical capital, intangible assets such as human capital, and technology may result in higher consumption, wages, and well-being in the future, but with the enormous risk to decrease the natural capital and, consequently, the biodiversity. Furthermore, erosion of the productive basis due to depreciation of assets, pollution, and depletion of natural resources may limit, or even reduce, future well-being. The implications of this perspective for the "global" economy are clear; to ensure future sustainability, the Hartwick rule (Hartwick, 1977) ought to be improved and technological progress, i.e., an increasingly intelligent use of existing assets, can play an immense role in future sustainability, especially taking into account the future development of the new economic giants of the 21st century (i.e., China and India).

References

Barbier, E., 2015. Nature and Wealth: Overcoming Environmental Scarcity and Inequality. Palgrave.

Barbier, E.B., 2011a. Capitalizing on Nature: Ecosystems as Natural Assets. Cambridge University Press.

Barbier, E.B., 2011b. Scarcity and Frontiers: How Economies Have Developed through Natural Resource Exploitation. Cambridge University Press, New York.

Barbier, E.B., 2019. How to Make the Next Green New Deal Work.

Barnett, H.J., Morse, C., 1963. Scarcity and Growth: The Economics of Natural Resource Availability. RFF Press.

Blum, M., Ducoing, C., McLaughlin, E., 2017. In: National Wealth: What Is Missing, Why it Matters. Oxford University Press, p. 89.

Climate Change Intergovernmental Panel, 2018. Global Warming of 1.5: An IPCC Special Report on the Impacts of Global Warming of 1.5 above Pre-industrial Levels and Related Global Greenhouse Gas Emission Pathways, in the Context of Strengthening the Global Response to the Threat of Climate Change, Sustainable Development, and Efforts to Eradicate Poverty. Intergovernmental Panel on Climate Change.

Coyle, D., 2015. GDP: A Brief but Affectionate History-Revised and Expanded Edition. Princeton University Press.

del Mar Rubio, M., 2004. The capital gains from trade are not enough: evidence from the environmental accounts of Venezuela and Mexico. Journal of Environmental Economics and Management 48 (3), 1175–1191.

Greasley, D., Hanley, N., McLaughlin, E., Oxley, L., 2016. Australia: A Land of Missed Opportunities?.

Hamilton, K., Clemens, M., May 1999. Genuine savings rates in developing countries. The World Bank Economic Review 13 (2), 333–356.

Hamilton, K., 1994. Green adjustments to GDP. Resources Policy, Elsevier 20 (3), 155–168.

Hanley, N., Dupuy, L., Mclaughlin, E., 2015. Genuine savings and sustainability. Journal of Economic Surveys 29 (4), 779–806.

Hartwick, J., 1977. Intergenerational equity and the investing of rents from exhaustible resources. The American Economic Review 67 (5), 972–974.

Hartwick, J.M., 1990. Natural resources, national accounting and economic depreciation. Journal of Public Economics 43 (3), 291–304.

Hartwick, J.M., 1991. Degradation of environmental capital and national accounting procedures. European Economic Review 35 (2–3), 642–649.

Helm, D., 2015. Natural Capital: Valuing the Planet. Yale University Press.

Helm, D., 2017. Sustainable Economic Growth and the Role of Natural Capital.

Jevons, W.S., 1866. The Coal Question: An Inquiry Concerning the Progress of the Nation, and the Probable Exhaustion of Our Coal-Mines. Macmillan, London.

Lindmark, M., 1998. Towards Environmental Historical National Accounts for Sweden (Ph. D. thesis).

Lindmark, M., Acar, S., 2013. Sustainability in the making? A historical estimate of Swedish sustainable and unsustainable development 1850–2000. Ecological Economics 86, 176–187.

Macedo, M.N., DeFries, R.S., Morton, D.C., Stickler, C.M., Galford, G.L., Shimabukuro, Y.E., 2012. Decoupling of deforestation and soy production in the Southern amazon during the late 2000s. Proceedings of the National Academy of Sciences 109 (4), 1341–1346.

Managi, S., Kumar, P., 2018. Inclusive Wealth Report 2018: Measuring Progress toward Sustainability. Routledge, New York, USA.

Meadows, D., Randers, J., 2012. The Limits to Growth: The 30-year Update. Routledge.

Pearce, D.W., Atkinson, G.D., 1993. Capital Theory and the Measurement of Sustainable Development: An Indicator of "Weak" Sustainability. Ecological Economics 8 (2), 103–108.

Pearce, D., 2002. An intellectual history of environmental economics. Annual Review of Energy and the Environment 27, 57–81.

Peskin, H.M., 1972. National Accounting and the Environment. Aschehoug.

Peskin, H.M., 1976. A national accounting framework for environmental assets. Journal of Environmental Economics and Management 2 (4), 255–262.

Stiglitz, J.E., Sen, A., Fitoussi, J.-P., 2009. Report by the commission on the measurement of economic performance and social progress. Sustainable Development 12, 292.

Withagen, C., Asheim, G.B., 1998. Characterizing sustainability: the converse of hartwick's rule. Journal of Economic Dynamics and Control 23 (1), 159–165.

World Commission on Environment and Development, 1987. Report of the World Commission on Environment and Development: Our Common Future (The Brundtland Report).

CHAPTER

Long waves and the sustainability transition

3

Mark Swilling

The crisis consists precisely in the fact that the old is dying and the new cannot be born; in this interregnum a great variety of morbid symptoms appear.
Antonio Gramsci

Introduction

The preamble to the Sustainable Development Goals (SDGs) refers to the need for a "transformed world." However, it says nothing about how this may come about or the historical trends that might coalesce to create the conditions for a transition to a transformed world. Following a tradition of thinking about "great transformations" pioneered by Polanyi (1946) and extended into the contemporary era by Berry (1999), this chapter will aim to address this challenge by proposing a framework for understanding the complex dynamics of sustainability transitions that are already underway. A growing body of popular and academic literature has turned to long-wave theory to contextualize the crisis and comprehend the dynamics of possible future trajectories of transition (for example, see Gore, 2010; Grin et al., 2010; Transitions et al., 2011; Swilling, 2013; Mason, 2015; Perez, 2016; Schot and Kanger, 2018). The problem with this literature is that the contributors tend to search for single long waves of transition (e.g., energy, or economic growth, or sociotechnological change) that are then used to better understand the wider dynamics of change. In reality, there are multiple historical dynamics unfolding at any one time that intersect in complex and unpredictable ways. This chapter breaks away from most long-wave traditions of scholarship by establishing a framework for making sense of the complexity of our current transition as a multiplicity of related, asynchronic and intersecting long-wave cycles.

It will be argued in this chapter that we need to understand the dynamics of the current global polycrisis as the emergent outcome of overlaps between four dimensions of transition: sociometabolic transition, technoeconomic transition, sociotechnical transition, and long-term global development cycles. When understood as

Handbook of Green Economics. https://doi.org/10.1016/B978-0-12-816635-2.00003-1

multiple cycles that intersect concurrently and asynchronistically across these four dimensions, the emergent outcome can be understood as a deep transition.[1] However, a deep transition is a quantum shift that must also—from a normative perspective—be a just transition. While an unjust sustainability transition is conceivable (decarbonized enclaves for rich elites and middle classes who want to be "green"), this will not provide a stable political basis for sustaining a deep transition over the long term. This normative perspective imposes additional challenges that are frequently avoided in most analyses of transition and resource efficiency.

These four dimensions of transition can combine into a deep transition in different ways depending on the dynamics and impacts of policy decisions and social struggles for more or less just outcomes. This goes to the *politics* of the deep transition. While there is agreement that we are in some kind of interregnum and that, therefore, a transition of some sort is almost inevitable, there is little agreement on how this transition should be understood. For some, exemplified by Perez (2002), 2009, 2013, the transition amounts to another phase in the evolution of industrial modernity with a reformed capitalism remaining as the primary mode of accumulation (for a similar position, see Mazzucato, 2016). In other words, it is assumed that industrial modernity is an epoch that began 250 years ago and that the transition is merely about coming into being of the next phase of this epoch, with some greening on the side where necessary. For others, we should anticipate a deep and fundamental transition to a postindustrial epoch of some kind (Schot and Kanger, 2018). This position bifurcates into two, with some arguing that a deep transition can only happen if corporate-dominated financialized capitalism is transcended (Korten, 1995; Eisenstein, 2013; Streeck, 2014; Escobar, 2015; Mason, 2015; Ahmed, 2017), while others assume that a post–industrial era is more or less synonymous with new more sustainable modes of production and consumption that remain essentially capitalist but in a very different form (Grin et al., 2010; Geels, 2013; Schot and Kanger, 2018). Either way, the debate about whether the outcome is capitalist or not can distract from the core issue, namely how a deep sociometabolic transition can also be a just transition. It cannot be assumed that the collapse of capitalism will automatically produce a more just outcome. Nor can it be assumed that the continuation of capitalism in its current form is consistent with a just transition. Capitalism takes many forms, and postcapitalism has yet to be clearly envisaged as more than a mere opposite of capitalism. What is certain is that if capitalism means private control of the large bulk of capital by a tiny fraction of the population (as documented by Picketty, 2014), this will be inconsistent with the requirements of a just transition. A deep and just transition creates conditions for alternatives, especially in the renewable energy sector.

[1] This term is borrowed from Schot and Kanger (see reference later in this chapter).

Rethinking the polycrisis from a long-wave perspective

The global economic crisis has generated a new literature that draws on long-wave theory to reimagine present and future landscapes. The writers in this neo-Polanyian tradition represent what Geels would refer to as clusters of discursive and cultural ontologies of probable futures (Geels, 2010). These include consultant's advisories and popular literature aimed at business audiences (Allianz Global Investors, 2010; Bradfield-Moody and Nogrady, 2010; Rifkin, 2011); the policy-oriented research-based literature generated from a variety of academic, UN, advisory, and consulting agencies (Von Weizsacker, et al. 2009; McKinsey Global Institute, 2011b; United Nations Environment Programm, 2011); the theory-laden academic literature (Drucker, 1993; Perez, 2009, 2010; Gore, 2010; Smith et al., 2010; Pearson and Foxon, 2012; Swilling and Annecke, 2012; Mason, 2015); and the postdevelopmental "transition discourses" (Escobar, 2015). These texts have all to a greater or lesser extent drawn on a tradition (originating in the works of Kondratieff and Schumpeter[2]) that depicts economic history in terms of a succession of long-term waves or cycles of economic development lasting between 40 and 60 years (for useful overviews of some of the main schools of long-wave—or what Foxon calls "coevolutionary"—thinking, see Foxon, 2011; Köhler, 2012).[3]

Acknowledging the critiques of long-wave theory (Broadberry, 2007) and the strong arguments in favor of coevolution and uneven development (Foxon, 2011), a framework will be proposed here that differs from existing approaches because it deals with four interactively asynchronous long-wave dynamics that operate at different temporal scales and with reference to four *dimensions of transition* (all explained in detail later):

- "sociometabolic transitions" that focus on the flow of materials and energy through socioecological systems across the preindustrial, industrial, and (potentially more sustainable) postindustrial epochs (Fischer-Kowalski and Haberl, 2007; Fischer-Kowalski and Swilling, 2011);
- "technoeconomic transitions"—or what Perez prefers to call "great surges of development"—comprising the evolution of the five main clusters of "general purpose technologies" (Lipsey et al., 2005) that have partially driven and shaped the fundamental changes in production and consumption during the industrial era (Perez, 2009);
- "sociotechnical transitions"—this refers to the multilevel perspective on the dynamics of change as "landscapes" interact with "regimes" and "niches" within particular sociotechnical sectors such as energy or transport (Geels, 2011);

[2] See Kondratief, 1935; Schumpeter, 1939.

[3] What is left out of this review are long-wave perspectives originating in evolutionary economics that do not include a reference to ecological cycles—a perspective originating in Nelson and Winter (1982) and expressed at a popular level in many references within business circles to supercycles—see report from global banking firm Standard Chartered (2010).

- "long-term global development cycles" is the Schumpeterian focus on cycles of economic growth, prices, crises, and "creative destruction" (Gore, 2010).

Polanyi was interested in a "double movement"—the divisive fragmentary nature of Laissez-faire capitalism and the integrative dynamics of microlevel pacts and associations—and proposed the coming of a "grand transformation." Social democracy after World War II (WWII) realized this grand prophecy. Today's conditions exhibit the same double movement—crisis, inequality, division, and potential collapse, versus the power of global grassroots movements expressing real livable alternatives. Today, there is increasing interest in the possibility of some sort of epochal shift, leading to a post–industrial world that is more or less sustainable—the "transformed world" referred to in the preamble to the SDGs. Schot and Kanger refer to this as the "second deep transition" (Schot and Kanger, 2018), whereas the German Advisory Council on Climate Change (explicitly invoking Polanyi) refers to another "great transformation" similar in significance to the agricultural and industrial revolutions 13,000 and 250 years ago, respectively (German Advisory Council on Climate Change, 2011). In this chapter, I will adopt Schot and Kanger's elegant notion of a "deep transition" broaden out its meaning to include the four dimensions of transition underpinned by an overarching normative commitment to a just transition.[4]

When integrated in this way, the following proposition becomes possible: the "deep transition" from industrial modernity to the "transformed world" referred to in the preamble of the SDGs is not merely about the extended survival of industrial modernity (embodied in its current financialized form of global capitalism with its various national manifestations), but rather it is about catalyzing a deep (sociometabolic) transition to a new sustainable epoch whose directionality and pace will depend on the three other dimensions of transition that emerge from within the contradictions of industrial modernity (technoeconomic transitions, sociotechnical transitions, and long-term development cycles). Depending on actually existing social struggles for change and how, in particular, the energy transition pans out, the outcome will be more or less *just.* A *just transition* may well need to be an information-based hybrid with capitalist features (e.g., a socially embedded market, subordination of finance to the "real economy," continuation of aspects of private ownership and private investment) and postcapitalist features (significantly expanded "commons" where ownership is neither state nor private, socially and/or publicly owned financial institutions with major investment resources, expanded nonmarket transactions, burgeoning social entrepreneurship sector, and a nonexploitative nonextractive relationship with natural systems). How exactly this pans out will more than likely be very different to what can be imagined from "this side of history" (Frase, 2016).

[4] In my review of Schot and Kanger's work, I proposed this broadening out of their concept, but my suggestion was not accepted for reasons that were not explained. The notion of a "deep transition" is thus used here to refer to what was called an "epochal transition" in previous work (Swilling and Annecke, 2012).

Sociometabolic transitions

There is an increasingly common trend within academic and nonacademic analyses of the crisis to identify purely economic causes of the crisis (of various kinds), followed by a set of conclusions about remedies that then add on at the end suggestions that the next phase of global growth will more than likely also be "green," "low carbon," or even "sustainable." This move amounts to an afterthought that recognizes the negative economic consequences of "externalities," but these externalities are left out of the analysis of the initial causes.[5] However, as Fischer-Kowalski points out, it is only possible to refer to the unsustainability of a system relative to another system (Fischer-Kowalski, 2011). To do this, she argues that the unit of analysis needs to be the sociometabolic flow of materials and energy through different configurations of coupled natural and social systems. The materials assessed usually include biomass, fossil fuels, construction minerals (mainly cement and building sand), and metals. These materials are measured in tons. By 2010, the global economy depended on the use of about 70 billion tons of resources per annum (Schandl et al., 2016). The reproduction of our current industrial civilization depends on a constant increase in the use of resources, plus the production of 500 exajoules of energy that generate the greenhouse gases that drive climate change. It is also an industrial civilization that is also highly unequal, with at most 20% of the population consuming over 80% of the worlds' resources and energy (United Nations Development Program, 1998).

Understanding socioecological systems in terms of the through-flow of materials from ecosystems through socioeconomic systems and back into ecosystems (as "waste") helps explain the deep transitions from one sociometabolic regime to another. Fischer-Kowalski and Haberl use this framework to reinterpret economic history since the last ice age. This includes deep transitions from hunter-gatherers to the agricultural sociometabolic epoch some 13,000 years ago as soils, seeds, and land became usable resources; from the agricultural to the industrial sociometabolic epoch over 250 years ago as fossil fuels, metals and minerals were added to the resource pool; and the "inevitable but improbable" (Fischer-Kowalski, 2011, p. 153) deep transition to a sustainable sociometabolic epoch when it is no longer possible to depend on large quantities of nonrenewable materials and cheap fossil fuels (Fischer-Kowalski and Haberl, 2007). By rooting the analysis of the polycrisis within the endogenous thermodynamics of material and energy flows, it becomes possible to anticipate futures where natural resources (and not just carbon) are used more efficiently and sustainably as a necessary condition for the emergence of a future potentially sustainable long-wave of ecologically sustainable and inclusive economic development.

[5] Of the literature cited thus far, the works by Allianz (2010) and Perez (2010b—including her contribution to this volume) are representative of this approach.

This perspective has been operationalized within the contemporary global policy space by the International Resource Panel (IRP) that was established in 2007 by (what is now called) United Nations Environment to deal with global material flows, resource depletion, and decoupling growth rates from rates of resource use (see http://www.unep.org/resourcepanel/) (Swilling, 2016).

The IRP distinguishes between four categories of resources: biomass, fossil fuels, construction materials, and metals.[6] In one of the IRP's first reports, the IRP advocated the controversial notion that well-being and "economic activity" (not just economic growth) could be decoupled from rising rates of resource use (Fischer-Kowalski and Swilling, 2011). Although this notion of decoupling is heavily criticized in the political ecology community (Jackson, 2009), what is not recognized in the literature is that this report has made a distinction between "relative decoupling" and "absolute resource reduction" irrespective of the prevailing growth rates and improvements in well-being. Yet, subsequent reports by the IRP did not follow through with this conceptual language, preferring rather to use the unfortunate distinction between "relative decoupling" and "absolute decoupling"—the former referring to conditions where the rate of growth in resource use is lower than the economic growth rate, and the latter referring to negative growth in resource use relative to the economic growth rate which, it is (incorrectly) assumed, needs to constantly rise. By hitching resource reduction to economic growth rates, the notion of "absolute decoupling" can unwittingly legitimize continued economic growth in economic regions where economic growth is not required (e.g., Europe and North America) (Ward et al., 2016). The alternative, which evolved out of a discussion with Bob Ayres, is to only refer to the distinction between the problematic notion of "relative decoupling" (economic growth plus *increases* in resource use, but at a slower rate than economic growth) and "absolute resource reduction." The latter is only possible if three conditions can be met: more resource efficiency (doing more with less) coupled to sufficiency (more for most who do not have enough, less for the overconsumers who make up 20% of the global population but consume over 80% of consumed goods), and a lot more is done with stuff that is not used extensively enough (i.e., renewables).

As the IRP reports show, rising global resource use during the course of the 20th century corresponded with declining real resource prices—a trend that came to an end in 2000–02. Since 2000–02, the macro trend in real resource prices has been upward (notwithstanding dips in 2008/2009 and in 2012).

The McKinsey Global Institute report (which was published after the IRP report) generally confirms the trends identified by the IRP report (McKinsey Global Institute, 2011a). The McKinsey report calculated that resource prices increased by 147% in the decade since 2000. Furthermore, up to $1.1 trillion is spent annually on what they call "resource subsidies." McKinsey argues that if resource subsidies

[6] Note that water and land resources are excluded from this categorization of global material flows—for a justification, see (Fischer-Kowalski and Swilling, 2011, p. 8–9).

are reduced, a carbon price of at least $30/ton is introduced and an additional $1 trillion per annum is invested in resource efficient production systems to meet growing demand, the result will be the creation of a whole new set of "productivity opportunities" with an internal rate of return of at least 10% at current prices. However, these are unlikely to become the focus for investments to drive economic recovery if resource subsidies continue to be defended by the institutionalized politically powerful interests of the dominant regimes of the financialized mineral-energy complex who vigorously defend "carbon lock-in" (Pierson, 2000).

The IRP's work essentially documents the resource limits of the industrial epoch and establishes empirically the rationale for a deep transition to a more sustainable epoch (Swilling, 2016). However, like the bulk of similar research in industrial ecology, ecosystem services, and climate change, a convincing case is made for *why* a deep transition is needed (mainly by accumulating vast amounts of quantitative analysis), but without any conception of *how* this will happen, and *what* needs to be achieved incrementally over the short, medium, and long term to ensure a sustainability-oriented deep transition actually takes place. In short, the IRP's work establishes the necessary conditions for a transition, but not the sufficient conditions. For this, it is necessary to focus on the complex dynamics of accumulation, institutional power, and technological change within the other three dimensions of transition.

Technoeconomic surges of development

The substantial body of work by Venezuelan economist Carlota Perez has deeply influenced those who write about technoeconomic cycles. Perez identified five technoeconomic transitions or what she preferred to call "great surges of development," each associated with specific technological revolutions that emerged at particular historic moments since the onset of the industrial revolution in the 1770s. Each followed the familiar S-curve with an installation and a deployment phase bifurcated by a financial crisis (Perez, 2002, 2007).

During the installation phase, financial capital (i.e., those who pursue capital gains by buying and selling shares in companies) tends to gravitate toward a particular cluster of promising sociotechnical innovations. Because investors have a "herd mentality," large amounts of money get concentrated in a selected interconnected set of innovations that, in turn, get catapulted into wider markets and aggressively marketed. This overexcited concentration of investments in innovations that must still be generally deployed within the wider economy and society leads to overinvestment, share price rises incommensurate with underlying value, and eventually, a financial bubble emerges. When this bubble bursts, a financial crisis (i.e., across-the-board devaluation of listed wealth that can, in turn, have wider economic repercussions) is triggered that requires (without implying this automatically happens) countercyclical state interventions to restore macroeconomic stability. This usually results in a shift in power from finance capital to productive capital (which is more long-termist and patient because it is dividend-seeking rather than capital gains oriented). After the 1929 crash, these state interventions were essentially inspired by Keynesian economic thinking, establishing the basis for the growth of the post-WWII welfare states. After the 1973 oil crisis, from the late 1970s onward, the interventions were essentially inspired by neoliberal economic thinking, establishing, in turn, the basis for neoliberal governance, financialization, and globalization during the 1980s and 1990s. Contrary to the rhetoric of neoliberalism, state intervention—and the increasingly significant role of the World Bank and IMF—was decisive in the global restructuring that occurred during this period.

A key problem with Perez's argument is that state interventions only really become significant in her analysis at times of crisis, and she underestimates the dynamics of capital accumulation. What she ignores is the increasingly significant body of evidence that states play a crucial role during the early phases of the innovation cycle with respect to two kinds of interventions (Mazzucato, 2015, 2016). The first is investments in R&D, and the second is investments and mechanisms to reduce risk during the early phases of the innovation cycle. These two make sense because in the case of the former, private investors are often reluctant to invest in R&D because the benefits accrue to society as a whole, not the specific investor. As for the latter, because risks tend to be too high for the private sector during the innovation cycle. Thus the "crowding in" of private investment depends on knowledge construction, risk reduction (via direct investments, or the provision of guarantees, or discounted loan finance), and the extent to which sunk costs have been depreciated. Mazzucato has applied this framework to explain both the rise of the Internet in the US context and the rise of renewables in the German and Chinese context (Mazzucato, 2011, 2015).

For Perez, there have thus been five major technoeconomic transitions, each corresponding with the emergence of a particular set of interdependent general-purpose sociotechnical innovations with transformative implications for regulatory regimes and institutional configurations. Each technoeconomic transition, however, goes through the S-curve that starts off with a finance-driven installation period and ends off with a productive capital-driven deployment phase, with the state playing key roles during the early phases of the innovation cycle and during the crisis-driven transition from finance to productive capital.

Following Perez (2002), the dominant technoeconomic paradigm today is what Castells referred to as the "Information Age" (Castells, 1997). It's origins lie in the United States in the early 1970s as innovators assembled the basics of which later became the microcomputing revolution that built on, but also transcended, the parameters of the fourth (fossil fuel–based) technoeconomic transition. However, Mason's reworking of this periodization makes more sense. Arguing against Perez's argument that the third transition lasted over 70 years (1908 to early 1970s), he prefers to date the fourth technoeconomic transition from 1945 to 2008 (i.e., corresponding with Gore's conception of the post-WWII long-term development cycle). The fourth is about the maturation of the fossil fuel era. For Mason, the fifth technoeconomic transition originates in the late 1990s (with the mainstreaming of

ICT) into many aspects of economic and daily life. For him and Perez, the crisis of 2007/2008 marked the midpoint of the fifth and the ending of the fourth technoeconomic transition. The oil crisis of the early 1970s marks the midpoint crisis of the fourth technoeconomic transition, not the start of the fifth as suggested by Mason (2015). While Mason's schema is more convincing, he does not recognize the significance of the rise of "green tech" and the RE revolution in the context of a gradual shift in power from finance to productive capital. "Greening" is not just an extension of the fifth technoeconomic transition. Instead, overlapping with the fifth technoeconomic transition (no matter whether one prefers Perez's or Mason's periodization) is the start of the sixth technoeconomic paradigm, triggered in part by the onset of the global financial crisis in 2007/2008 and reflected in the brief appearance of "green Keynesianism" during 2009 (Geels, 2013). The best example of this is the massive escalation in investment in renewable energy from around US$159 billion in 2007 to US$280 billion in 2017, with even steeper increases in public sector investment in renewables (from US$2 billion in 2004 to around US$70 billion by 2014) than in private sector investment (from just US$20 billion in 2004 to nearly US$80 billion by 2014) (Mazzucato and Semieniuk, 2018). Public sector investments have underwritten and catalyzed private sector investment in renewables. Thus, breaking with both Perez and Mason, it can be argued that the sixth technoeconomic transition commenced in the early 2000s that is probably best described as the "green tech" revolution, with renewable energy as the lead sector.

Perez has argued that the global economic crisis of the "Information Age" (the fifth technoeconomic transition) has, in fact, experienced a "double bubble"—the so-called "dot.com" bubble of 1997–2000, followed by the financial bubble of 2004–07. The former was triggered by overinvestments in the so-called "dot.com" shares during the years leading up to the "dot.com" crash in 2000, and the latter by overinvestments in financial instruments. Perez has argued that these "two bubbles of the turn of the century are two stages of the same phenomenon" (Perez, 2009, p. 780), i.e., together they mark the midpoint crisis of the fifth technoeconomic transition (which is, of course, the Information Age, whether or not it is seen as originating in the 1970s as per Perez or the 1990s as per Mason). She argues against the Keynesian argument that explains the financial crisis as a "Minsky moment" in terms of which debt markets have an in-built tendency toward financial instability, which can only be mitigated by increased state spending after the crisis-driven devaluation hits (Krugman, 2012). Instead, she argues that the most significant crises are triggered by the financial opportunities created by new technologies that result in "major technology bubbles" that eventually burst. This is what the "Internet mania" of 1997–2000 was all about. However, instead of triggering an economic recession that would have necessitated extensive state intervention to prepare the way for productive capital to take over from financial capital after the bubble burst in 2000/2001, the postcrisis recession was mitigated by the rapid financialization of the global economy, the rise of China as the manufacturer and de facto funder of the world economy, and the accelerated expansion of information-based global trade (with flexible specialization and just-in-time systems as key operating procedures).

Cheap Chinese exports (achieved in part by "artificially" keeping the value of the Chinese yuan down) not only brought down the cost of mass consumer goods (which effectively raised the value of wages in the west at a time when real wages were flat or in decline), but they also made it possible for China to become one of the world's largest providers of debt to developed world consumers via the purchase of massive quantities of government bonds (which were then on-loaned to the private Banks). Indeed, the preference for liquid assets and quick operations within the paper economy that Chinese surpluses made possible generated skyrocketing capital gains for finance capital in the lead up to the dot.com crisis between 1996 and 2000, whereas profits in the real economy (i.e., sectors outside of IT) remained flat or even negative. After the "dot com" crash, instead of interventions to restrain financial capital, the opposite happened as various interventions by the Federal Reserve and neoliberal governments around the world effectively allowed the paper economy to mushroom into a gigantic unregulated global casino (Gowan, 2009; Turner, 2016)—what former UK Prime Minister Gordon Brown liked to call "light touch" regulation.

For Perez, a key condition for a successful transition is the disciplining of capital gain-seeking finance capital to make way for dividend-seeking productive capital to drive the deployment phase. However, in her writing (Perez, 2013), she has started to factor in environmental externalities by emphasizing the role that innovations for greening the economy will play in the deployment phase of the fifth cycle. But it is unclear what will drive these innovations—finance or productive capital? Nor does Perez define a new historic mission for finance capital after it has been disciplined to make way for productive capital. Indeed, there is little evidence that finance capital is in fact being effectively disciplined (see discussion below). It will be suggested that a solution to this problem lies in accepting that the sixth cycle—the "green tech revolution"—may be emerging that is increasingly driven by finance capital supported by state interventions to generate R&D and reduce risk for investors. Surely this is the new historic mission for finance capital, especially during a period of falling RE prices and awareness of the potential threats of climate change. And would this not create a growth-catalyzing installation phase of the emergent sixth cycle to complement the deployment phase of the fifth? Perez is reluctant to accept this line of argumentation, as is Mason. Yet this may well be what is underway.

Sociotechnical transitions

Although the sociotechnical literature is rooted in the broader literature on systems innovation, evolutionary economics, and the sociology of technology (Rip and Kemp, 1998), for the purpose of this chapter I use the multilevel perspective (MLP) on sociotechnical transitions (Geels, 2005; Smith et al., 2010; for a critique from a political ecology perspective, see Lawhon and Murphy, 2011). According to the MLP, sociotechnical transitions result in "deep structural changes" over long time periods within particular sectors (e.g., transport, energy, water, sanitation, waste communications) that involve fundamental reconfigurations of technologies, markets, institutions, knowledge, consumption practices, and cultural norms (Geels, 2011, p. 24). They are explained in terms of complex nondeterministic interrelations between three levels of reality: landscape pressures (macro), regime structures (meso), and niche innovations (micro). This framework is then used to address the challenge of the complex transition(s) to a more sustainable world, which is defined as "human well-being in the face of real biophysical limits" (Meadowcroft, 2011, p. 71) and "an open-ended orientation for change" (Grin et al., 2010, p. 2).

Major sociotechnical regimes comprise a core set of technologies that coevolve with social functions, social interests, market dynamics, policy frameworks, and institutional regulations. These sociotechnical regimes are shaped by a broad constituency of technologists, engineers, policymakers, business interests, NGOs, consumers, and so on. The interrelationships of these interests through regulations, policy priorities, consumption patterns, and investment decisions, among other things, hold together to stabilize sociotechnical regimes and their existing trajectories. Regimes set the parameters for what is possible:

> *... reconfiguration processes do not occur easily, because the elements in a socio-technical configuration are linked and aligned to each other. Radically new technologies have a hard time to break through, because regulations, infrastructure, user practices, maintenance networks are aligned to the existing technology...*
>
> **Geels (2002), p. 1258.**

The concept of "landscape" is important in the MLP in seeking to understand the broader "conditions," "environment," and "pressures" for transitions. The landscape operates at the macrolevel beyond the immediate efficacy of human agency, focusing on issues such as political cultures, economic growth, macroeconomic trends, land use, utility infrastructures, and so on. The landscape applies pressures on existing sociotechnical regimes creating opportunities for responses, for example, climate change and the need for the expansion of renewable energy (RE). Landscapes are characterized as being "external" pressures that have the potential to impinge upon—but do not determine—the constitution of regimes and niches: they are an external context "... that sustains action and makes some actions easier than others. These external landscape developments do not mechanically impact niches and regimes, but need to be perceived and translated by actors to exert influence..." (Geels and Schot, 2007, p. 404).

The idea of sociotechnical niches, which operate at a microlevel, is one of "protected" spaces, usually encompassing small networks of actors learning about new and novel technologies and their uses. These networks agitate to get new technologies onto "the agenda" and promote innovations by trying to keep alive novel technological developments. The constitution of networks and the expectations of a technology they present are important in the creation of niches.

The MLP is particularly useful when it comes to explaining why particular sectors introduce sustainability-oriented alternatives, from green packaging (retail sector), lighter materials (car industry), recycling (waste industry), mass transit (mobility sector), organic foods (food system), and renewable energy (energy sector). In each case, it is relatively easy to demonstrate how a preexisting regime became unviable as a result of the impact of landscape pressures (climate change, prices of renewables, water scarcities, urbanization, etc). However, for change to occur, niche innovations need to mature to a point where they can become an alternative regime or else get absorbed by incumbent regimes that decide to survive by transitioning (e.g., some of the large European energy companies that have become major renewable energy players). Once again, this cycle unfolds over time, normally a 40–60 year period.

Global development cycles

To improve our understanding of the linkages between the technoeconomic transitions that Perez has identified and the sociotechnical transitions that emerge from the MLP (with special reference to the RE revolution) and the long-wave dynamics of global economic development, it is necessary to turn to the work of UNCTAD economist Charles Gore (Gore, 2010). He has located the technoeconomic cycles described by Perez within what he refers to as the Kondratieff-like "global development cycle" that began in the 1950s and ended with the global economic contraction of 2009. For Gore, the year 2009 marks a key turning point because it was the only year since 2009 that the global economy actually shrank. For Gore, a Kondratieff-type cycle cannot be equated to the technoeconomic cycles that Perez has in mind. While technoeconomic cycles typically follow the well-known S-curve (found in Perez and to some extent the MLP) of "*irruption-crisis-deployment*", as Fig. 3.1 suggests the global development cycles adhere to very different logics. The global development cycle starts off with "*growth-plus-price-inflation*" during the spring-summer period (1950s/1960s) ending in a stagflation crisis driven in part by overinvestment in infrastructures during the growth phases (1970s). These overinvestments push up interest rates prior to the benefits of the infrastructure investments working their way through the economy as a whole. This is then followed by *growth-with-limited-inflation* during the autumn-winter period (1980s/1990s) ending in deflationary depression driven in part by diminishing returns on mature technologies, while returns on the new technologies have yet to materialize and the inflationary pressures have not kicked in yet (2007 onward). Significantly, the first half of the post-WWII long-term development cycle was dominated in the West by a Keynesian "golden age" of welfarism, inclusion, solidarity, and liberation (including decolonization in the peripheries) within national development policy frameworks, whereas the second was dominated by neoliberalism, exclusion, structural adjustment, commodification, individualism, and rising inequalities in an increasingly globalized world.

Although Perez tried to link technological cycles to economic growth, in her later work she gave up this effort. Although Gore (2010) admits that there is no evidence to support the notion that growth phases are driven exclusively by technoeconomic

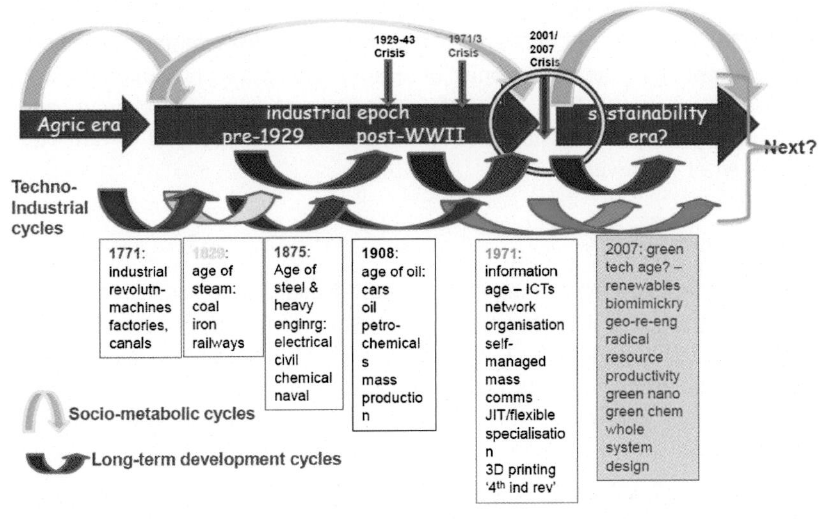

FIGURE 3.1

Sociometabolic cycles, long-term development cycles, and technoindustrial cycles, 1770s–post 2009.

Source: author.

transitions he has enriched the overall picture by correlating the price cycles derived from Berry (1991) with the technoeconomic cycles derived from Perez's work. Read together, these are very rough approximations of actual growth cycles without in any way suggesting that the actual complex drivers of economic growth at any moment in time are reducible to these long-wave dynamics. His key insight seems to confirm Köhler's argument that S-curves do not run consecutively (as represented both by the MLP and, to some extent, by Perez), but instead they tend to overlap with the deployment phase of a previous cycle and the installation of the new cycle acting as codrivers of growth-oriented processes (Köhler, 2012).

Specifically, Gore argues that the post-1970s growth phase was driven by *two* key drivers: the *first* was the economics of the deployment phase of the fourth technoeconomic transition (investments in fossil fuels, mass production, inclusive infrastructures, global trade) and the *second* was the installation phase of the fifth technoeconomic transition (information and communication technologies and their associated applications). The midpoint crisis (2007/2009) of the fifth technoeconomic transition (Information Age) also marks the *end* of the post-WWII global development cycle that ended in 2009. Like the interregnum between 1929 and the 1940s, an interregnum between 2007/2009 and the take-off of the next long-term development cycle can be expected. However, the next global long-term development cycle could only emerge if radical institutional reconfigurations not only displace finance capital to unleash productive capital (following Perez's script) but also displace the powerful and highly subsidized regimes of the mineral-energy complex that depend on the continuities of "carbon lock-in" (Pierson, 2000). A key factor that prevents this shift is the persistent epistemological attachment within economics to the reductionist obsession with prices and markets, a crude one-dimensional materialism, a belief in the rational *homo economicus*, and the extraordinarily simplistic assumption that market economies tend toward equilibrium instead of the reverse.

Toward a synthesis

There is significant evidence that, over the past decade, there has been a significant increase in investments in renewable energy, communications, and mobility (Oxford Economics, 2015; Baller et al., 2016; United Nations Environment Program, 2018).These empirical realities are significant because they provide the required focus for convincingly connecting the four dimensions of transition. Consider the following statements:

- the commencement of the "spring period" of a long-term development cycle (or Kondratieff cycle) usually corresponds with significant increases in investment in three foundational infrastructures, namely energy, communications, and mobility (World Bank, 2017): the upticks in investments in RE (mainly finance capital, with state subsidies) (see United Nations Environment Program, 2018), mass transit (productive capital, with a big role for DFIs) (Oxford Economics, 2015), and Web 3.0 (productive capital, mainly the newly consolidated giant IT companies) (Baller et al., 2016) tend to confirm this, with the last providing the operating system for the first two (as argued by Rifkin, 2011);
- the commencement of a Perezian technoeconomic surge corresponds with the increases in investments by finance capital in new technoeconomic innovations: the move by finance capital into "green tech" (especially RE) suggests that the installation phase of the sixth technoeconomic transition has begun (United Nations Environment Program, 2018);
- a regime shift within a particular sector that results in a sociotechnical transition (as per the MLP) corresponds with the coalescing of niche innovations into a new regime that is more appropriately aligned with landscape pressures: the RE revolution follows the ideal-typical dynamic suggested by the MLP's sectoral focus with niches forming into regimes that emerge in parallel to the old fossil fuel energy regimes (Geels et al., 2017), but there is also evidence of old regimes absorbing niches as in the case of large fossil fuel—based energy companies who change their business models to include—or even prioritize—RE (for example, Enel, Italy's largest energy company) (Aklin and Urpelainen, 2018);
- Energy Return on Investment a sociometabolic transition is underway when there is a fundamental shift in resource use patterns, often taking many years to occur: the drastic decline in the EROI ratio (from 1:100 in the 1930s to 1:10 today) and the rise of renewables seem to suggest that such a "civilizational transition"

(Ahmed, 2017) may already be starting, and if what is unfolding is a resource price supercycle, this could well accelerate the decarbonizing dynamics of the sociometabolic transition.

In short, the increased investments in energy, communications, and mobility confirm there is evidence of transitional dynamics across the four dimensions of transition. This provides the basis for concluding that a deep transition of some kind has commenced.

However, under what conditions will this be a just transition? If it is not also a just transition, it can be a deep transition that could well be destabilized and even attenuated by social movements and political backlashes.

A just transition—as opposed to a mere transition to a decarbonized economy—may well depend on state institutions (such as the new generation of Development Finance Institutions [DFIs] and Sovereign Wealth Funds [SWFs]) and large nonprofit financial institutions (such as Triodos Bank, or the German Church banks) stepping in to create appropriate "financial commons" for channeling dividend-oriented long-termist investments into social enterprises. There is a growing body of literature on "energy democracy" that argues that the material reality of distributed decentralized renewable energy infrastructures favors the building of this kind of decentralized community-based financial ecosystem (Burke and Stephens, 2018). The example of Germany stands out: around 50% of the renewable energy operations were socially or publicly owned by 2012 (Debor, 2018). This is the kind of institutional-cum-financial innovation that will make it possible for a new generation of tech-savvy social enterprises to emerge across a wide range of different contexts, from rural electrification in East Africa to the advanced so-called "smart grid" apps in the developed world that could potentially create the operating system for new modes of energy democracy.

Conclusion

Much depends on whether a new interconnected cluster of low-carbon "general-purpose technologies" and associated institutional configurations (especially in the finance sector) will emerge to drive the next long-term development cycle. In other words, can the sociotechnical transition underway in the energy sector become a binding force that catalyzes new sets of interconnected value chains across all other mainstream sectors. Pearson and Foxon have concluded that contrary to much contemporary optimism about the potential for a low-carbon industrial transition, there is little evidence of this (Pearson and Foxon, 2012).[7] However, like all publications of this nature prior to 2016, such pessimism may be outdated. The acceleration in the uptake of renewables in most world regions is truly remarkable

[7] For a contrary view, see Jänicke (2012)

(Aklin and Urpelainen, 2018), making references from before 2016 almost redundant. What this suggests is that finance capital has started to discover its historic mission as the driver of the installation phase of the sixth technoeconomic transition, often taking advantage of the R&D constructed with state subsidies and risk-reducing state interventions (Mazzucato, 2015). However, as suggested by the German case where 50% of renewable energy enterprises were communally/socially owned by 2012, there may well be a new generation of financial institutions that combine the familiar role of backing new disruptive innovations with a just transition agenda within the wider framework of the new "commons." This is reinforced by the rise of the sharing and "zero marginal cost" economy that Mason regards as the entry point into postcapitalism (Mason, 2015). The kinds of infrastructures that become the focus of mainstream investment flows over the next two decades, however, will provide the clues as to what kind of wider deep transition may be unfolding. As argued elsewhere, a key space to watch is investments in urban infrastructures because towns/cities are rapidly emerging as the fulcrums of the next potentially more sustainable global development cycle (Hodson et al., 2012). Where these dovetail with new RE investments, the potential for energy democracies becomes an exciting set of possibilities (Davies et al., 2017).

It remains doubtful, however, that conditions have matured to a point where the present interregnum can be transcended in a way that could result in a more inclusive and sustainable long-term development cycle. While there is some debate about whether the low-carbon and resource efficiency technologies have matured sufficiently (Janicke, 2012; vs. Pearson and Foxon, 2012), what is becoming very clear is that the consolidation—through a spate of mergers and acquisitions—within the information and communication sector is preparing the way for the deployment phase of the Information Age. With a strategic focus on "digitization" and "integrated value chains," the conditions are in place for productive capital to take the lead (Acker et al., 2012). However, many analysts (including Perez and Gore) admit that this time round it might not be so easy to discipline financial capital to make way for productive capital. In this respect the Marxists have a point when they argue that the structural nature of contemporary global capitalism is such that finance capital has managed to establish a hegemonic role for itself that may allow it to resist the transition to productive capital (Altvater, 2009; Gowan, 2009; Harvey, 2009; Blackburn, 2011). However, this argument will weaken if the practical Keynesianism of the type expressed in the Stiglitz Report on how to restructure the global financial system without dismantling capitalism manages to be implemented as part of a radical shake-up of the global financial power structures (Stiglitz, 2010) or if regulatory measures to "fix" global finance are implemented (Turner, 2016). It may also weaken if the energy democracy movement spurns new modes of socially responsible "green finance," especially if these "commons" are reinforced by the DFIs and SWFs.

For many observers, by 2018, there seemed little evidence of any fundamental restructuring of the global financial system, thus confirming Gore's original argument that the global crisis is a "structurally blocked transition" (Gore, 2010). We have the rivalry between China and the United States about the value of the Chinese currency and trade barriers; the ongoing financial instabilities in the EU, exacerbated by the multiple sovereign debt crises; the de facto bankruptcy of the United States masked by "Quantitative Easing" (read: printing money) and the (short-term) fracking bonanza; the relatively unfettered flow of speculative finance through global markets despite the Dodd-Frank regulatory reforms in the United States (which the Trump Administration has now dismantled); the hoarding of cash as investors wait for short-term capital gain opportunities to return, instead of looking for long-term productive investments in the real economy; the continued expansion of derivatives and the power of hedge funds; the uptick in investments in fossil fuels since the commencement of the Trump Presidency; and national governments who, having experienced massive devaluations in the past, continue to build up currency reserves to counteract financial shocks, thus keeping much-needed investment capital away from productive investment. The increased indebtedness of the United States, as it cuts taxes and increases spending, has most likely triggered a trajectory that could culminate in the next financial crash. However, there are also countermovements, including the rise of the DFIs and SWFs as key dividend-oriented long-term investors, the rise of social enterprises (e.g., in the renewables sector in Germany) and associated socially responsible financial institutions, the shift into "green tech" by increasing numbers of investment funds, the creation of new global funds for mitigating climate change, the rising significance of infrastructure investment funds focused on urban infrastructure, and the increasing number of major businesses who are committing to more sustainable practices.

We do, indeed, live in a highly complex interregnum when—as Gramsci noted about a previous era—the old seems to be dying and the new is struggling to be born. While a great variety of "morbid symptoms" may be appearing, many hopeful signs are also evident.

References

Acker, O., Grone, F., Schroder, G., 2012. The Global ICT 50: The Supply Side of Digitization. Strategy and Business, Autumn, pp. 52–63.

Ahmed, N.M., 2017. Failing States, Collapsing Systems: Biophysical Triggers of Political Violence. Springer, Cham, Switzerland.

Aklin, M., Urpelainen, J., 2018. Renewables: The Politics of a Global Energy Transition. MIT Press, Boston.

Allianz Global Investors, 2010. The Sixth Kondratieff - Long Waves of Prosperity. Allianz Global Investors, Frankfurt. Available at: http://www.allianzglobalinvestors.de/capitalmarketanalysis. [Accessed: 1 March 2012].

Altvater, E., 2009. Postneoliberalism or Postcapitalism? the Failure of Neoliberalism in the Financial Market Crisis. Development Dialogue, January.

Baller, S., Dutta, S., Lanvin, B., 2016. The Global Information Technology Report 2016: Innovating in the Digital Economy, WEF, Insead. World Economic Forum, Geneva. https://doi.org/10.17349/jmc117310.

Berry, B.J.L., 1991. Long-wave Rhythms in Economic Development and Political Behaviour. John Hopkins University Press, Baltimore and London.
Berry, T., 1999. The Great Work: Our Way into the Future. Bell Tower, New York.
Blackburn, R., 2011. 'Crisis 2.0', New Left Review, 72(Nov/Dec), pp. 33–62.
Bradfield-Moody, J., Nogrady, B.T., 2010. The Sixth Wave: How to Succeed in a Resource-Limited World. Vintage Books, Sydney.
Broadberry, S., 2007. Recent Developments in the Theory of Very Long Run Growth: A Historical Appraisal. Warwick University, Department of Economics Research Papers, Warwick.
Burke, M.J., Stephens, J.C., 2018. Political power and renewable energy futures: a critical review. Energy Research and Social Science 35, 78–93.
Castells, M., 1997. The Information Age Volumes 1, 2 and 3. Blackwell, Oxford.
Davies, M., Swilling, M., Wlokas, H.L., 2017. Towards new configurations of urban energy governance in South Africa's renewable energy procurement programme. Energy Research and Social Science. https://doi.org/10.1016/j.erss.2017.11.010.
Debor, S., 2018. Multiplying Mighty Davids: The Influence of Energy Cooperatives on Germany's Energy Transition. Springer, New York.
Drucker, P., 1993. Post-capitalist Society. Butterworth-Heinemann, Oxford.
Eisenstein, C., 2013. The More Beautiful World Our Hearts Know Is Possible. North Atlantic Books, Berkeley.
Escobar, A., 2015. Degrowth, postdevelopment, and transitions: a preliminary conversation. Sustainability Science 10 (3), 451–462. https://doi.org/10.1007/s11625-015-0297-5.
Fischer-Kowalski, M., 2011. Analysing sustainability transitions as a shift between socio-metabolic regimes. Environmental Transition and Societal Transitions 1, 152–159.
Fischer-Kowalski, M., Haberl, H., 2007. Socioecological Transitions and Global Change: Trajectories of Social Metabolism and Land Use. Edward Elgar, Cheltenham, U.K.
Fischer-Kowalski, M., Swilling, M., 2011. Decoupling Natural Resource Use and Environmental Impacts from Economic Growth, Report for the International Resource Panel. United Nations Environment Programme, Paris.
Foxon, T.J., 2011. A coevolutionary framework for analysing a transition to a sustainable low carbon economy. Ecological Economics 70 (12), 2258–2267. https://doi.org/10.1016/j.ecolecon.2011.07.014.
Frase, P., 2016. Four Futures. Verso, London.
Geels, F., 2011. The multi-level perspective on sustainability transitions: responses to seven criticisms. Environmental Innovation and Societal Transitions 1 (1), 24–40.
Geels, F., 2013. The impact of the financial-economic crisis on sustainability transitions: financial investment, governance and public discourse. Environmental Innovation and Societal Transitions 6, 67–95.
Geels, F.W., 2002. Technological transitions as evolutionary reconfiguration processes: a multi-level perspective and a case-study. Research Policy 31 (8/9), 1257–1274.
Geels, F.W., 2005. Technological Transitions: A Co-evolutionary and Socio-Technical Analysis. Edward Elgar, Cheltenham, UK.
Geels, F.W., 2010. Ontologies, socio-technical transitions (to sustainability), and the multi-level perspective. Research Policy 39 (4), 495–510.
Geels, F.W., Sovacool, B.K., Schwanen, T., Sorrell, S., 2017. Sociotechnical transitions for deep decarbonization. Science 357 (6357), 1242–1244. https://doi.org/10.1126/science.aao3760.

Geels, F.W., Schot, J., 2007. Typology of sociotechnical transition pathways. Research Policy 36 (3), 399–417. https://doi.org/10.1016/j.respol.2007.01.003.

German Advisory Council on Climate Change, 2011. World in Transition: A Social Contract for Sustainability. German Advisory Council on Global Change, Berlin.

Gore, C., 2010. Global recession of 2009 in a long-term development perspective. Journal of International Development 22, 714–738.

Gowan, P., 2009. Crisis in the heartland. New Left Review 55, 5–29.

Grin, J., Rotmans, J., Schot, J., Geels, F., Loorbach, D., 2010. Transitions to Sustainable Development: New Directions in the Study of Long Term Transformative Change. Routledge, New York.

Harvey, D., 2009. The Crisis and the Consolidation of Class Power: Is This Really the End of Neoliberalism? Counterpunch.

Hodson, M., Marvin, S., Robinson, B., Swilling, M., 2012. Reshaping urban infrastructure: material flow analysis and transitions analysis in an urban context. Journal of Industrial Ecology 16 (6). https://doi.org/10.1111/j.1530-9290.2012.00559.x.

Jackson, T., 2009. Prosperity without Growth? the Transition to a Sustainable Economy. Sustainable Development Commission, United Kingdom.

Janicke, M., 2012. "Green Growth": from a growing eco-industry to economic sustainability. Energy Policy 48, 13–21.

Köhler, J., 2012. A comparison of the neo-Schumpeterian theory of Kondratiev waves and the multi-level perspective on transitions. Environmental Innovation and Societal Transitions 3, 1–15.

Kondratief, N.D., 1935. The long waves in economic life. Review of Economic Statistics 27 (1), 105–115.

Korten, D., 1995. The Post-Corporate World. Berret-Koehler, San Francisco.

Krugman, P., 2012. End This Depression Now. W.W. Northon & Company, New York & London.

Lawhon, M., Murphy, J.T., 2011. Socio-technical regimes and sustainability transitions: insights from political ecology. Progress in Human Geography. https://doi.org/10.1177/0309132511427960.

Lipsey, R.G., Carlaw, K.I., Bekar, C.T., 2005. Economic Transformations: General Purpose Technologies and Long Term Economic Growth. MIT Press, Cambridge, MA.

Mason, P., 2015. Postcapitalism: A Guide to Our Future. Penguin, London.

Mazzucato, M., 2011. The Entrepreneurial State. Demos, London.

Mazzucato, M., 2015. The green entrepreneurial state. In: Scoones, I., Leach, M., Newell, P. (Eds.), The Politics of Green Transformations. Routledge Earthscan, London and New York, pp. 133–152.

Mazzucato, M., 2016. Rethinking Capitalism: Economics and Policy for Sustainable and Inclusive Growth. John Wiley & Sons, West Sussex.

Mazzucato, M., Semieniuk, G., 2018. Financing renewable energy: who is financing what and why it matters. Technological Forecasting and Social Change 127, 8–22.

McKinsey Global Institute, 2011a. Resource Revolution: Meeting the World's Energy, Materials, Food, and Water Needs. McKinsey Global Institute, London. Available at: http://www.mckinsey.com/client_service/sustainability.aspx.

McKinsey Global Institute, 2011b. Urban World: Mapping the Economic Power of Cities. McKinsey Global Institute, London.

Meadowcroft, J., 2011. Engaging with the politics of sustainability transitions. Environmental Innovation and Societal Transitions 1, 70–75.

Nelson, R.R., Winter, S.G., 1982. An Evolutionary Theory of Economic Change. Harvard University Press, Cambridge, MA. OECD, 2011. Towards Green Growth. OECD, Paris.

Oxford Economics, 2015. Assessing the Global Transport Infrastructure Market: Outlook to 2025, PwC. PriceWaterhouseCoopers, London. Available at: www.pwc.com/outlook2025.

Pearson, P.J.G., Foxon, T.J., 2012. A low carbon industrial revolution: insights and challenges from past technological and economics transformations. Energy Policy 50, 117–127.

Perez, C., 2002. Technological Revolutions and Financial Capital: The Dynamics of Bubbles and Golden Ages. Elgar, Cheltenham, U.K.

Perez, C., 2007. Great surges of development and alternative forms of globalization. In: Kattel, R. (Ed.), Working Papers in Technology Governance and Economic Dynamics, Norway and Estonia: The Other Canon Foundation (Norway) and Tallinin University of Technology (Tallinin).

Perez, C., 2009. The double bubble at the turn of the century: technological roots and structural implications. Cambridge Journal of Economics 33, 779–805.

Perez, C., 2010. The financial crisis and the future of innovation: a view of technical change with the aid of history. In: Working Papers in Technology Governance and Economic Dynamics. Norway and Tallinin. The Other Canon Foundation & Tallinin University of Technology.

Perez, C., 2013. Unleashing a golden age after the financial collapse: drawing lessons from history. Environmental Innovation and Societal Transitions (0). https://doi.org/10.1016/j.eist.2012.12.004.

Perez, C., 2016. Capitalism, Technology and a Green Golden Age: The Role of History in Helping to Shape the Future. WP, 2016-1.

Picketty, T., 2014. Capital in the Twenty-First Century. Belknap Press, Boston, MA.

Pierson, P., 2000. Increasing returns, path dependence, and the study of politics. American Political Science Review 94 (2), 251–267.

Polanyi, K., 1946. Origins of Our Time: The Great Transformation. Victor Gollancz Ltd, London.

Rifkin, J., 2011. The Third Industrial Revolution: How Lateral Power Is Transforming Energy, the Economy and the World. Palgrave MacMillan, New York.

Rip, A., Kemp, R., 1998. Technological change. In: Rayner, S., Malone, E.L. (Eds.), Human Choice and Climate Change. Oxford University Press, Oxford, pp. 327–399.

Schandl, H., Fischer-Kowalski, M., West, J., Giljum, S., Dittrich, M., et al., 2016. Global material flows and resource productivity. Assessment Report for the UNEP International Resource Panel. https://doi.org/10.1111/jiec.12626.

Schot, J., Kanger, L., 2018. Deep transitions: emergence, acceleration, stabilization and directionality. Research Policy 46 (7), 1045–1059. https://doi.org/10.1016/j.respol.2018.03.009.

Smith, A., Voss, J.P., Grin, J., 2010. Innovation studies and sustainability transitions: the allure of the multi-level perspective and its challenges. Research Policy 39 (4), 435–448.

Standard Chartered, 2010. The Super-Cycle Report. Global Research Standard Chartered, London. Available: Gerard.Lyons@sc.com [accessed March 2012].

Stiglitz, J.E., 2010. The Stiglitz Report: Reforming the International Monetary and Financial Systems in the Wake of the Global Crisis. New Press, New York.

Streeck, W., 2014. How will capitalism end? New Left Review 87 (june), 35–64, 9781784784034.

Swilling, M., 2013. Economic crisis, long waves and the sustainability transition: an African perspective. Environmental Innovation and Societal Transitions (0). https://doi.org/10.1016/j.eist.2012.11.001.

Swilling, M., 2016. Preparing for global transition: implications of the work of the international resource Panel. Handbook on Sustainability Transition and Sustainable Peace 378–390. https://doi.org/10.1002/sd.244.

Swilling, M., Annecke, E., 2012. Just Transitions: Explorations of Sustainability in an Unfair World. United Nations University Press, Tokyo.

Transitions, J., World, U., Town, C., 2011. From: Swilling, M. & Annecke, E. 2011. Just Transitions: Explorations of Sustainability in an Unfair World. Cape Town: Juta, pp. 1–49.

Turner, A., 2016. Between Debt and the Devil: Money, Credit and Fixing Global Finance. Princeton University Press, Princeton.

United Nations Development Programme, 1998. Human Development Report 1998. United Nations Development Programme, New York, 5 November 2006.

United Nations Environment Programm, 2011. World Economic and Social Survey 2011: The Great Green Technological Transformation. World Economic and Social Survey, New York: United Nations.

United Nations Environment Programme, 2018. Global Trends in Renewable Energy Investment 2018. United Nations Environment Programme, Nairobi. Available at: http://fs-unep-centre.org/publications/global-trends-renewable-energy-investment-report-2018.

Von Weizsacker, E., Hargroves, K.C., Smith, M.H., Desha, C., Stasinopoulos, P., 2009. Factor Five: Transforming the Global Economy through 80% Improvements in Resource Productivity. Earthscan, London.

Ward, J., Sutton, P., Werner, A., Costanza, R., Mohr, S., Simmons, C., 2016. Is decoupling GDP growth from environmental impact Possible ? PLoS One 1–14. https://doi.org/10.1371/journal.pone.0164733. https://doi.org/DOI:10.137.

World Bank, 2017. Investments in IDA Countries: Private Participation in Infrastructure. World Bank, Washing D.C.

CHAPTER

4 Greening of industry in a resource- and environment-constrained world

Izzet Ari, Riza Fikret Yikmaz

Introduction

Environment and natural resources are under stress due to rapid industrialization, urbanization, and population increase combined with unsustainable consumption patterns. Economic development and underlying patterns of unsustainable production and consumption are the main drivers of the environmental problems such as climate change, air pollution, loss of biodiversity, and water and soil degradation. In addition, changes in the industrial production and consumption patterns, driven by technological advances, have significantly increased the use of natural resources and chemical compounds and at the same time increased the wastes. In responding these issues, sustainability considerations have taken more attention in the international arena, and many contesting approaches have developed for the transformation of economies into more sustainable and greener ones.

Sustainability has first been introduced into the global political agenda in 1972 by the UN Conference on the Human Environment, also known as Stockholm Conference (UN, 1972). In 1987, this concept then evolved into "Sustainable Development" by the World Commission on Environment and Development who defined it as "(the) development that meets the needs of the present, without compromising the ability of future generations to meet their own needs" (WCED, 1987). In the UN Conference on Environment and Development (UNCED) in Rio de Janeiro in 1992, it became a global agenda through Agenda 21, which opened a new era, and sustainable development became overarching goal of international community. In 2012, 20 years after Rio Conference, in the UN Conference on Sustainable Development (Rio+20), political ambition on sustainable development is renewed by decision on the development of Sustainable Development Goals (SDGs) and adoption of guidelines on green economy. In 2015, the 2030 Agenda for Sustainable Development (UN, 2015) was adopted at the UN Conference on Sustainable Development. The 2030 Agenda features 17 SDGs as shown in Fig. 4.1 and 169 associated targets, which are built on the Millennium Development Goals (MDGs) adopted in 2000.

Handbook of Green Economics. https://doi.org/10.1016/B978-0-12-816635-2.00004-3

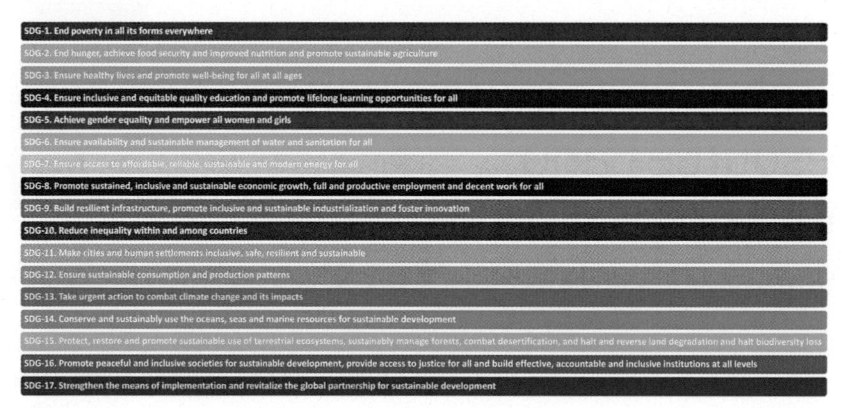

FIGURE 4.1

Sustainable development goals.

Credit: Prepared by Authors, data from UN, 2015. Transforming Our World: The 2030 Agenda for Sustainable Development. Retrieved December 21, 2015, from: https://sustainabledevelopment.un.org/post2015/transformingourworld

Together with these international developments, the economic crises and environmental problems such as the threat of climate change triggered the elaboration of existing political and economic agendas and lead to the consideration of green economy or green growth, circular economy, and sustainable consumption and production (SCP) in the context of sustainable development.

In this study, fundamental points for development of policies for green industries are highlighted, and well-known linkages among circular economy, SCP, and greening industries are presented. Sustainable Development Goals provide the main framework for this analysis. Green industrial policies and practices are considered with SCP and circular economy model to achieve SDGs as a whole.

Conceptual framework

Green economy, or interchangeably, green growth, is defined by international organizations such as Organisation for Economic Co-operation and Development (OECD) and United Nations Environment Programme (UNEP) as an approach prioritizing the investment and consumption of the goods and services that support environmental improvements. This is introduced as a catalyzing concept to support achievement of sustainable development. This concept has received significant international acceptance over the last decade as a tool to address the financial crisis in 2008 and was adopted as one of the two themes for the Rio+20 Conference in 2012 (UN, 2018). At Rio+20, the green economy is considered to be inclusive and able to drive economic growth, employment, and poverty eradication, while maintaining the sustainability of environment. Improving capacity, exchange of information, and sharing of experience are given importance for implementing green economy policies in the outcome document of the Rio+20.

Sustainable consumption and production

The need for SCP first appeared on the international agenda in 1992 UN Conference on Environment and Development held in Rio de Janeiro. At this conference, the major cause of the deterioration of the global environment was accepted as the unsustainable pattern of consumption and production. In 1994, the first definition of SCP was proposed at the Oslo Symposium on Sustainable Consumption as "*the use of services and related products, which respond to basic needs and bring a better quality of life while minimizing the use of natural resources and toxic materials as well as the emissions of waste and pollutants over the life cycle of the service or product so as not to jeopardize the needs of further generations*" (Ofstad et al., 1994). In 2002, SCP was recognized in the Johannesburg Plan of Implementation, one of the outputs of the World Summit on Sustainable Development. The Johannesburg Plan of Implementation emphasized the need for development of a 10-year framework of programs in support of the regional and national initiatives to accelerate the shift toward SCP. In 2015, SCP became one of the most important components of the 2030 Agenda for Sustainable Development. Twelfth SDG of the Agenda aims to ensure SCP patterns.

SCP is a concept of fostering resource and energy efficiency, sustainable infrastructure, and access to basic services and providing green and decent jobs and a better quality of life for all. It is a holistic approach that minimizes the negative environmental impacts of consumption and production systems while promoting quality of life for all (Akenji et al., 2015). From the production side, SCP refers to a set of cleaner production practices and increasing resource efficiency, which are enabled by innovation and technological change. On the consumption side, SCP implies changing the consumption patterns of households and governments through changes in lifestyles and individual consumer behavior and choices, as well as through changes in procurement strategies in the public sector (World Bank, 2017b).

Circular economy

Circular economy is a recently defined phenomenon for greening industries. Similar to other sustainable development models, circular economy addresses decoupling, resource efficiency, production efficiency, slower material flows rather than linear economic models, and lower resources extraction without reducing economic activity (Mccarthy et al., 2018). Besides, circular economy is stated as a continuous positive development cycle that conserves and enriches natural capital, optimizes resource yields, and reduces system risks by managing finite stocks and renewable flows. A circular economy is invigorating and recovering by design and targets to keep products, components, and materials at their highest utility and value at all times (EU, 2017). Long-lasting design, maintenance, repair, reuse, remanufacturing, refurbishing, and recycling are identified as the ways to achieve circular economy (Camilleri, 2018). New opportunities such as emerging sectors based on secondary material production remanufacturing, reducing risks on supply security from imported materials, and creating new decent jobs are identified as the major potential advantages of the circular economy practices (Mccarthy et al., 2018). Existing literature argues that circular economy has positive impact on economic growth and job creation (Mccarthy et al., 2018).

Greening industries

Greening industry, as a strategy, addresses unsustainable production and consumption patterns to achieve green growth and green economy in manufacturing sectors and resource scarce areas such as energy, forests, land, water, and food (United Nations Secretary-General's High-Level Panel on Global Sustainability, 2012). Besides, the greening of industry includes environmental performance that monitors resource efficiency, degree of absolute or relative decoupling (that is, the rate of gross domestic product [GDP] growth higher than the rate of resource extraction or pollutants, greenhouse gas [GHG] emissions), reducing impacts of industrial activities on environmental goods, natural resources, and materials (Schwarzer, 2013). Policies for greening industries are complements of the principles and policies of SCP and circular economy. Although greening industries are narrowed on manufacturing sectors, they nevertheless aim to promote SCP patterns from raw materials to final consumer products and encourage to follow reduce, reuse, and recycle all the materials on earth (Stoddart et al., 2011). Greening industries require national, regional, and local green industry policies with political ownership and well-defined framework of actions (UNIDO, 2011). UNIDO Green Industry: Policies for Supporting Green Industries (UNIDO, 2011) lists policy framework, support and initiatives, and science and technologies for greening industries. Apart from the United Nations Industrial Development Organization (UNIDO), other international organizations such as the World Bank, the OECD, and the UN and scholars focus on conceptual framework for green industrial development.

Greening industries need political commitments including national sustainable development strategies, national SCP strategies, and national development plans. All these plans and strategies can explicitly focus on concrete industrial policies in terms of dimensions of sustainable development and can also integrate these policies into other sectors. Besides, these policies are required to achieve vertical integration to regional and local policy documents (UNIDO, 2015). Policies for greening industries are supported by some tools such as cleaner production, life cycle management, ecolabeling, and environmental accounting. These tools are designed either as part of legislation or as voluntary agreements among governments, private sector, civil society, and consumers. Implementation of green industry policies and initiatives requires involvement of all stakeholders to be progressive. Although the public sector, particularly government, line ministries, and local authorities are responsible for policymaking and designing in general, research institutes, private sector, and civil societies provide capacity building in science and technology, and create financial resources for green policies and practices (OECD, 2012).

In addition to wide participation, capacity building and training for these stakeholders are essential to sustain green industry practices. Building capacity in all segments of society provides new opportunities such as generating innovations, creating research and development ecosystem, dissemination and deployment of greener technologies, emerging new financial resources to greener practices, and building new partnerships such as clusters, symbiosis, science parks, joint ventures, and incubators to expand knowledge and experiences (UNIDO, 2011).

Green industry policy requires cost-effective solutions to internalize externalities such as GHG emissions, pollutants such as waste, air, and water pollution. Market-based instruments such as tax, trade, and charges provide sectoral and technological transformation as well. For example, putting a price on carbon enables to minimize overall cost of GHG emission reductions and change consumption and production patterns (Stern, 2007). In response to imposing a price on carbon, producers try to use more efficient production mechanisms and eliminate or reduce emissions through shifting away carbon intense or material intense mechanisms. Carbon pricing can also create revenue for implementation of greener policies. Besides carbon case, other externalities such as wastes, air pollutants, overextraction of natural resources, unsustainable products, and hazardous chemicals can be subject to pricing to evolve conventional production patterns.

Science and technology complement policy frameworks for green industrial development. Technological progress is the engine of green economy (UN, 2011). The number of countries that engage in technological innovation for transition to greener industries is increasing to take comparative advantages of new growth model (The World Bank, 2012). Energy sectors, particularly renewable energy technologies, are well-known cases for development of greener technologies. Unlike other commodities in economy, increasing demand for renewable energy technologies leads to higher production of these technologies both in developed and in developing countries and alleviates total costs of renewable energy technologies (UN, 2011). Dissemination of renewable energy technologies in rural parts of developing countries provides inclusive and sustainable development (UNIDO, 2015), decouples economic growth, and increases economic productivity (Schwarzer, 2013; UN, 2011). In addition to producing renewable energy technologies in manufacturing industries, energy efficiency, cleaner production, waste minimization, and recovery increase overall resource efficiency and productivity in industries (Schwarzer, 2013). Nevertheless, new and best available technologies in various sectors are required to achieve sustainable development.

Greening industries and Agenda 2030

The industry-related goals and targets of Agenda 2030 are analyzed by UNIDO for creating shared prosperity, advancing economic competitiveness, and safeguarding the environment (UNIDO, 2015). There are 169 SDG targets, 131 of which are directly related to inclusive and sustainable industrial development. 65 targets are substantially or partially related to industry. 19 targets are listed as irrelevant for industry. Fig. 4.2 presents relevance of targets for inclusive and sustainable industrial development. SDG-1 "End poverty in all its forms everywhere" and SDG-9 "Build resilient infrastructure, promote inclusive and sustainable industrialization and foster innovation" have the highest relevance ratio as 100%. SDG-8 (Decent Work and Economic Growth) and SDG-12 (SCP) have 92% direct relevance. SDG-5: Gender Equality (56%), SDG-6: Clean Water and Sanitation (75%), SDG-7: Affordable and Clean Energy (80%), SDG-13: Climate Action (60%), SDG-14: Life Below Water (70%), and SDG-17: Means of Implementation (74%) have targets, which are very explicitly related to green industry. For example, SDG-1 (End Poverty) aims to promote small- and medium-sized enterprises with creating new jobs, increasing prosperity, and better environmental and health conditions. This goal is benefitting from inclusive and sustainable industrial development.

SDG-2 (Zero Hunger) includes actions within agroindustry and agrobusiness for food security, new jobs, creating income, and building linkage between farmers and industries. Therefore, half of the targets are directly related to industrial development (UNIDO, 2015). SDG-3 (Good Health and Well-being) is benefitting from

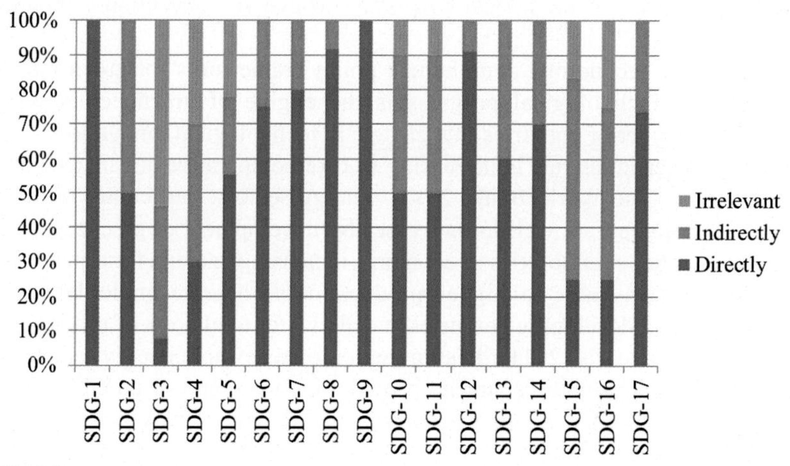

FIGURE 4.2

Inclusive and sustainable industrial development.

Credit: Prepared by Authors, data from UNIDO, 2015. The 2030 Agenda for Sustainable Development: Achieving the Industry-related Goals and Targets. Retrieved from: https://www.unido.org/fileadmin/user_media_upgrade/Who_we_are/Mission/ISID_SDG_brochure_final.pdf

industrial development through innovation, research and development (R&D) in medical treatments and technologies, and vaccines for all people. Similarly, SDG-10 (Reduced Inequalities) addresses all kinds of inequalities such as economic, gender, and vulnerable demographics. Industrial development enables environment to tackle inequalities among and within countries either developed or developing, accelerates structural transformation and social mobility, provides new jobs and additional income, and reduces social exclusion (UNIDO, 2015).

SDG-11 (Sustainable Cities) and industrialization are hand in hand. Cities are the places where infrastructure, industry, economic activity, innovation, resource management, environmental, and social issues are considered together. This goal aims to reduce impacts of industries on natural resources, environment, and climate change. Besides, this goal provides better environment for increasing effectiveness of environmental services and goods (UNIDO, 2015). Likewise, SDG-6 (Clean Water and Sanitation) is highly related to sustainable industries because industries require water for their activities and efficient use of water in industrial processes is important to lower the impacts on environment. SDG-14 (Life Below Water) is close to SDG-6 at this point. Oceans, seas, and marine ecosystems are affected by conventional industrial activities. Reducing marine pollution and ocean acidification is targeted by SDG-14. Sustainable industrial practices are required for conserving marine areas.

SDG-4 (Quality Education), SDG-5 (Gender Equality), and SDG-16 (Peace, Justice and Strong Institutions) are related to skills, equal participation in economic activities, and peaceful and inclusive societies, respectively. Industries need well-educated skills and capable societies for production, innovation, and technological transformation. Therefore, quality education is essential for sustainable industrialization. Empowering women and girls can increase overall productivity of societies. Inclusive and sustainable industrial development requires stability without conflict, extreme poverty, lack of security, violence, and corruption (UNIDO, 2015). SDG-4, 5, and 16 enable environment for better industrialization.

SDG-7 (Affordable and Clean Energy) and SDG-13 (Climate Action) both affect and are affected by industrial development. To shift to low-carbon development, energy efficiency and renewable energy are two main areas. Green industries provide technological and process interventions for this transition (UNIDO, 2015). GHG emission reduction can also be ensured through sustainable industrial development. Unsustainable industrial processes account for 33% of global total CO_2 emissions, cause to environmental and natural resources degradation, pollution, climate change, and negative externalities (UNIDO, 2015). It is essential that the unsustainable industrial activities should be transformed to greener and low-carbon industries. Mitigation, adaptation, technology transfer, finance, and capacity building are the main elements of climate policy (Ari, 2017). Industrial activities are linked with all these elements for tackling climate change. Climate-friendly practices, technologies, and processes in industries could minimize the impacts of climate change, reduce GHG emissions, provide new patterns for capacity building in societies, and create new and additional finance for business. Similarly, SDG-15 (Life on Land) aims to reduce impacts of agricultural production, deforestation, urbanization, and climate change due to unsustainable industrial activities.

SDG-8 (Decent Work and Economic Growth), which has strong linkage between industrial development, addresses economic growth with increasing productivity for structural transformation in industries (UNIDO, 2015). In the same way, SDG-9 (Industry, Innovation and Infrastructure) highlights that sustained economic growth and sustainable development requires sustainable and inclusive industrialization. SDG-17 (Means of Implementation), in turn, includes tools such as finance, technology transfer, trade, partnership and cooperation, and systemic issues to realize SDGs. Industries have a role in all these tools to change both consumption and production patterns. SDG-12 (SCP) addresses sustainable and efficient production patterns, efficient resource management, circular economy, cleaner production, waste minimization, etc. This goal promotes sustainability certification for industry and monitoring environmental performance of industries (UNIDO, 2015). To move toward low-carbon development and SCP patterns, "green industries," which create link between economic growth and environmental goods and services, can play an important role providing new opportunities such as decent jobs, income, and prosperity (UNIDO, 2015).

Being specifically set for achieving SCP, SDG-12, which consists of 11 targets and 13 indicators, should also be highlighted here. The key themes of SDG-12 are the efficient use of natural resources, reduction of food waste and food loss, management of chemicals and all wastes through life cycle, reduction of waste generation, and rationalization of inefficient fossil fuel subsidies (Box 4.1).

There are several SCP-linked targets placed under different SDGs related to agriculture, water, energy, and economic growth. More specifically, target 2.4 under SDG-2 (End hunger, achieve food security and improve nutrition, and promote sustainable agriculture), target 6.4 under SDG-6 (Ensure availability and sustainable management of water and sanitation for all), targets 7.2 and 7.3 under SDG-7 (Ensure access to affordable, reliable, sustainable and modern energy for all), and target 8.4 under SDG-8 (Promote sustained, inclusive and sustainable economic growth, full and productive employment and decent work for all) have close interlinks with respect to achieving SCP (World Bank, 2017b).

SDG target 1.4 aims to ensure equal rights to economic resources, including natural resources, so its scope is linked with efficient use of natural resources. SDG targets 2.3 and 2.4 on zero hunger and food security focus on both resource efficiency and sustainable agriculture practices, and these targets are directly linked with SCP. Food waste and loss are much diversified among countries, so SDGs emphasize this issue in target 12.3. Food production and consumption patterns are a concern for sustainability in food supply chain including food production, management, logistics, and consumption. While food waste and loss are high in some developed countries in terms of their consumption behavior, they are also significant in developing and least developed countries due to insufficient management and logistics (World Bank, 2018).

SDG targets 3.9 and 3.c focus on healthy lives and well-being. The former focuses on environmental health management including waste and pollution prevention, whereas the latter aims to education and training for healthy workforce to

Box 4.1 SDG-12 targets and key themes (in bold) (UN, 2015)

Goal 12. Ensure sustainable consumption and production patterns

12.1 Implement the 10-year framework of programmes on sustainable consumption and production, all countries taking action, with developed countries taking the lead, taking into account the development and capabilities of developing countries.
12.2 By 2030, achieve the sustainable management and efficient use of natural resources.
12.3 By 2030, halve per capita global food waste at the retail and consumer levels and reduce food losses along production and supply chains, including post-harvest losses.
12.4 By 2020, achieve the environmentally sound management of chemicals and all wastes throughout their life cycle, in accordance with agreed international frameworks, and significantly reduce their release to air, water and soil in order to minimize their adverse impacts on human health and the environment.
12.5 By 2030, substantially reduce waste generation through prevention, reduction, recycling and reuse.
12.6 Encourage companies, especially large and transnational companies, to adopt sustainable practices and to integrate sustainability information into their reporting cycle.
12.7 Promote public procurement practices that are sustainable, in accordance with national policies and priorities.
12.8 By 2030, ensure that people everywhere have the relevant information and awareness for sustainable development and lifestyles in harmony with nature.
12.a Support developing countries to strengthen their scientific and technological capacity to move toward more sustainable patterns of consumption and production.
12.b Develop and implement tools to monitor sustainable development impacts for sustainable tourism that creates jobs and promotes local culture and products.
12.c Rationalize inefficient fossil fuel subsidies that encourage wasteful consumption by removing market distortions, in accordance with national circumstances, including by restructuring taxation and phasing out those harmful subsidies, where they exist, to reflect their environmental impacts, taking fully into account the specific needs and conditions of developing countries and minimizing the possible adverse impacts on their development in a manner that protects the poor and the affected communities.

increase production efficiency. Therefore, these targets are substantially related to SCP. Target 4.7 ensures that all learners acquire the knowledge and skills needed to promote sustainable development, including sustainable lifestyles.

Fig. 4.3 presents material footprint, which attributes global material extraction for using domestic final demand for countries. This figure shows that high-income countries consume more materials than middle- and low-income countries. In line with the target 12.2 namely sustainable management and efficient use of natural resources, many developed countries and emerging economies need to reduce their material footprint through following SCP policies. In this respect, SDG target 6.4 aims to increase water use efficiency and address water scarcity, target 7.2 aims to increase share of renewable energy, and target 7.3 aims to improve energy efficiency, where all are directly associated with SCP. SDG target 8.4, which aims improving resource efficiency in consumption and production to decouple economic growth from environmental degradation, is directly linked with SCP. Target 8.9,

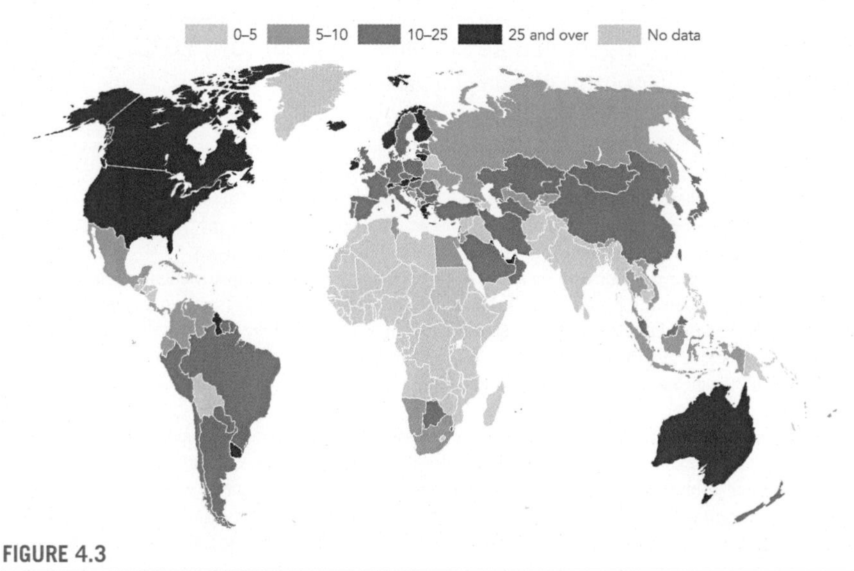

FIGURE 4.3

Material footprint, 2010 (metric tons per capita).

Credit: From World Bank, 2018. Atlas of Sustainable Development Goals (2018): World Development Indicators. Washington, DC. https://doi.org/10.1596/978-1-4648-1250-7. License: Creative Commons Attribution CC BY 3.0 IGO.

which focuses on sustainable tourism that creates jobs and promotes local culture and products, is indirectly related to SCP.

SDG target 9.4 focuses on retrofitting industries to make them sustainable together with increased resource use efficiency and greater adoption of clean and environmentally sound technologies and industrial processes, and is regarded to be very likely to contribute to deployment of SCP policies. Similarly, target 9.5 on enhancing scientific research upgrades the technological capabilities of industrial sectors through encouraging innovation and research and development and substantially provides input to more SCP practices. Industrial diversification in medium- and high-tech industries, including chemical and machinery manufacturing industries, depends on their technological facilities, innovation capacity, and employment (World Bank, 2018). The indicator, which is used for evaluation of countries' industrialization level and share of industries in the whole economy, is the manufacturing value added in GDP (Fig. 4.4). This share is the highest in North America, European region, China, and India. The lowest share is observed in sub-Saharan Africa (World Bank, 2017a).

Energy intensity refers to energy used to produce one dollar's out. This indicator reflects not only energy efficiency of the whole economy but also energy intensities of the industrial sectors. SDG-7, 9, and 13 are exclusively related to energy intensity of industries. Especially, SDG target 7.3 aims to double the global rate of improvement in energy efficiency. This target is one of the drivers in changing SCP patterns

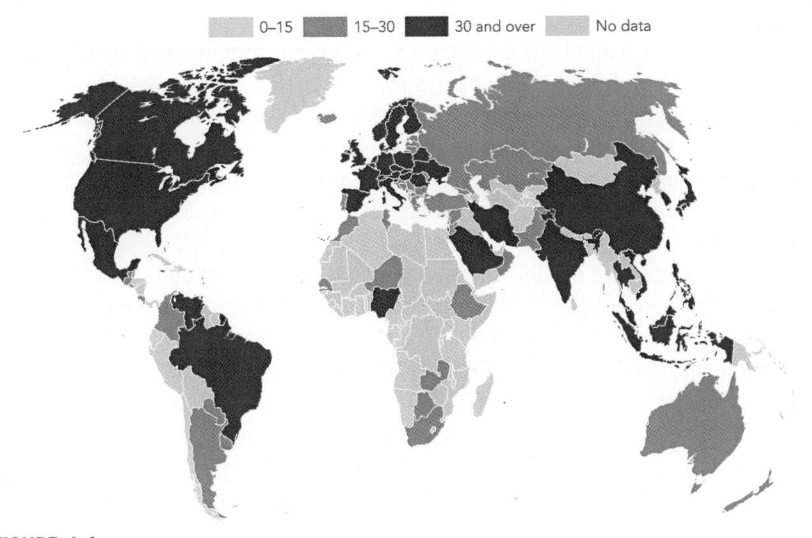

FIGURE 4.4

Medium- and high-tech industry (% manufacturing value added).

Credit: From World Bank, 2018. Atlas of Sustainable Development Goals (2018): World Development Indicators. Washington, DC. https://doi.org/10.1596/978-1-4648-1250-7. License: Creative Commons Attribution CC BY 3.0 IGO.

through increasing energy efficiency, decreasing energy intensity, and declining carbon emissions. Nevertheless, absolute decoupling has not been observed at the global level yet. Although energy intensity has been declining at global level, it is still far in the Middle East and North Africa, and the global industrial production remains highly energy intensive (World Bank, 2018). The global average of energy intensity was 5.1 (MJ/2011 PPP $ GDP) in 2015, and it is a significant progress when comparing 1990 level as 7.6 (World Bank, 2018). However, this decline is just for high-income countries and a few emerging economies.

SDG target 11.2 provides sustainable transport systems, 11.3 enhances inclusive and sustainable human settlement planning and management, 11.6 reduces environmental impact of cities, including air pollution waste, and 11.b encourages implementing integrated policies and plans including resource efficiency, which can be listed as examples of SCP practices in urban areas. Similarly, SDG target 13.2 aims to integrate climate change measures into national policies, strategies, and planning and target 13.3 encourages to improve education, awareness raising, and human and institutional capacity on climate change, where both are substantially linked with SCP.

SDG target 14.1 prevents marine pollution, 14.4 regulates harvesting and end overfishing, 14.6 prohibits overfishing, and 14.7 increases sustainable use of marine resources, including through sustainable management of fisheries, aquaculture, and tourism. Consequently, four targets in SDG-14 are directly linked with SCP in marine, sea, and ocean activities. SDG target 15.1 ensures sustainable use of the freshwater ecosystems and their services, 15.2 promotes implementation of sustainable management of forests, and 15.a and 15.b increase financial resources to conserve and sustainably use biodiversity, forests, and ecosystems. Hence, all these targets ensure SCP in life on land activities. Besides, SDG target 16.6 develops effective, accountable, and transparent institutions to achieve SDGs. Therefore, review and follow-up of the SDGs, particularly 10-year framework on SCP, is open to all evaluation.

Table 4.1 summarizes the linkages between SDG-12 (SCP) and other SDGs. 40 targets are directly related to SCP. Target on sustainable management and efficient use of natural resources, namely SDG-12.2, has the highest number of linkages (16 linkages) among other targets in SDG-12. SDG-9 (Industry, Innovation and Infrastructure) is closer to SCP than other SDGs. Therefore, the connection between SCP and greening of industries is observed in the set of SDGs.

Conclusion

Economic growth, urbanization, industrialization, and population growth will likely have adverse consequences on natural resources, environment, and human well-beings. Beginning with the Stockholm Conference in 1972, the subject of sustainability has been discussed among governments, international organizations, private and public sectors, civil societies, academia and research institutes, financial organizations, and individuals. The document "Our Common Future Report" (1987) is a critical milestone in the cross area of environment-economy-society for defining "Sustainable Development" concept. The Earth Summit in 1992 brought sustainable development principles and comprehensive multilateral environmental agreements: the United Nations Framework Convention on Climate Change and the United Nations Convention on Biological Diversity (UN, 1992). In 2000, MDGs were set as eight goals to measure progresses on sustainable development among developing countries. In 2012, as a 20-year anniversary of the Earth Summit in 1992, the Rio+20 UN Conference on Sustainable Development was held. This conference resulted in the report "The Future We Want," which recognizes the preparation of SDGs. Between 2013 and 2015, stakeholders including countries, civil societies, private sectors, and academia worked on developing SDGs. In 2015, the SDGs were adopted under the UN General Assembly. Thus, these chronological milestones briefly show the evolution of sustainable development in the world agenda.

During this journey, many concepts for decoupling environmental abatement from economic growth had been initiated, such as pollution prevention and control, life cycle assessment, environmentally friendly products, waste minimization, reduce-reuse-recovery-recycle, green economy, green growth, SCP, circular economy, inclusive and sustainable growth, low-carbon economy, and low-carbon development. All these terms, models, and approaches have the common motivation that is called sustainable

Table 4.1 Interlinkages between SCP and other SDGs, prepared by authors.

	12.1	12.2	12.3	12.4	12.5	12.6	12.7	12.8	12.a	12.b	12.c
1.4		✓									
2.3		✓	✓								
2.4		✓	✓						✓		
3.9					✓						
3.c				✓							
4.7								✓			
6.4		✓									
7.2		✓									✓
7.3		✓									✓
8.4	✓										
8.9										✓	
9.4	✓	✓		✓					✓		
9.5									✓		
11.2				✓							
11.3				✓							
11.6				✓							
11.b	✓	✓					✓				
13.2	✓										
13.3								✓			
14.1		✓		✓							
14.4		✓									
14.6		✓									
14.7		✓									
15.1		✓									
15.2		✓									
15.a		✓									
15.b		✓									
16.b						✓	✓				

development. Besides, all these are based on reducing impacts of human activities on natural resources, environment, and planetary boundaries on earth.

These concepts are all interconnected and aim at achieving sustainability of natural resources through resource and energy efficiency. For instance, green economy or green growth is an economy-wide and inclusive approach, encompassing social aspects of life such as sustain decent jobs and poverty eradication. In fact, SCP could be considered as a natural element of the green economy. Although closely related to SCP, circular economy is a comparatively narrower concept focusing on industrial processes to minimize wastes considering life cycle of products.

All of these concepts have transformative impacts on industries around the world. Encouraging cleaner production, decoupling pollutants from economic outputs, pollution control, reducing the waste and emissions in industrial processes, eco-efficiency, increasing the efficiency in energy, water and raw material use, and resource recovery are the common goals. Decoupling, cleaner production, eco-efficiency, resource and material efficiency, reducing emissions, and pollutants can be defined not only as initiatives or practices but also as the key indicators to measure progresses on achieving sustainable development. Energy, manufacturing industries, agriculture, and transportation are the main sectors to be monitored for sustainable development. These sectors are related to each other so changing production and consumption patterns in one of them can trigger the progress on achieving sustainable development.

In this study, green industrial transition is analyzed within a policy perspective. The concept of greening industries requires a comprehensive policy framework, political commitments and dedicated actors, well-designed initiatives and instruments, strong collaboration with science and technology, partnerships among public, private, and academia, finance, technology transfer, and capacity building including training and education. These entire requirements provide a guide through analyzing relationship between industry and sustainable development in the frame of SDGs and SCP. Among 169 SDG targets, 131 of them are directly related to industry, 65 targets are substantially or partially related, and 19 targets have no relevance. Further analysis is based on relation between SCP and targets of SDGs. Targets of SDG-12 (SCP) are followed within other goals. The number of directly related targets is 40. SDG-12 has 11 targets, and other SDGs have 40 targets, which are directly linked with SCP. SDG-9 (Industry, Innovation and Infrastructure) has the highest relevance with SDG-12, so this shows that SCP and green industrial policies are designed hand in hand.

In conclusion, SDGs can be a golden opportunity to guide all sectors to transform their policies and practices. Green industrial policies should continue to evolve within the Agenda 2030 period to achieve solid patterns of sustainable development.

References

Akenji, L., Bengtsson, M., Briggs, E., Chiu, A., Daconto, G., Fadeeva, Z., Tabucanon, M., 2015. Sustainable Consumption and Production: A Handbook for Policymakers. United Nations Environment Programme. https://doi.org/10.13140/2.1.4203.8569.

Ari, I., 2017. The interconnection between sustainable development goals and the climate change negotiations: the paris agreement case. Alternatif Politika (Special Issue), 27–45. Retrieved from. http://alternatifpolitika.com/site/cilt/9/sayi/0/2-Ari-Climate-Change.pdf.

Camilleri, M.A., 2018. Closing the loop for resource efficiency, sustainable consumption and production: a critical review of the circular economy. Forthcoming International Journal of Sustainable Development (2013), 1–22. Retrieved from. https://ssrn.com/abstract=3119575.

EU, 2017. Workshop Report: Promoting Remanufacturing, Refurbishment, Repair and Direct Reuse. https://doi.org/10.2779/181003.

Mccarthy, A., Dellink, R., Bibas, R., 2018. The Macroeconomics of the Circular Economy Transition: A Critical Review of Modelling Approaches. https://doi.org/https://doi.org/10.1787/af983f9a-en.

OECD, 2012. Greening Development: Enhancing Capacity for Environmental Management and Governance. OECD Publishing. https://doi.org/10.1787/9789264167896-en.

Ofstad, S., Westly, L., Bratelli, T., 1994. Symposium: Sustainable Consumption. Ministry of Environment, Oslo, Norway.

Schwarzer, J., 2013. Industrial Policy for a Green Economy (IISD Report). Retrieved from: https://www.iisd.org/sites/default/files/publications/industrial_policy_green_economy.pdf.

Stern, N., 2007. Carbon pricing and emissions markets in practice. In: The Economics of Climate Change, vol. 32. Cambridge University Press, Cambridge, pp. 368–392. https://doi.org/10.1017/CBO9780511817434.025.

Stoddart, H., Riddlestone, S., Vilela, M., 2011. Principles for the Green Economy. Retrieved from. http://earthcharter.org/virtual-library2/principles-for-the-green-economy/.

The World Bank, 2012. Inclusive Green Growth. The World Bank. https://doi.org/10.1596/978-0-8213-9551-6.

UN, 1972. Report of the United Nations Conference on the Human Environment. Stockholm.

UN, 1992. The Earth Summit. Retrieved July 2, 2019, from. https://www.un.org/geninfo/bp/enviro.html.

UN, 2011. The Green Economy: Trade and Sustainable Development Implications. Geneva.

UN, 2015. Transforming Our World: The 2030 Agenda for Sustainable Development. Retrieved December 21, 2015, from. https://sustainabledevelopment.un.org/post2015/transformingourworld.

UN, 2018. Green Economy: Sustainable Development Knowledge Platform. Retrieved September 8, 2018, from. https://sustainabledevelopment.un.org/topics/greeneconomy.

UNIDO, 2011. Policies for Supporting Green Industry.

UNIDO, 2015. The 2030 Agenda for Sustainable Development: Achieving the Industry-Related Goals and Targets. Retrieved from. https://www.unido.org/fileadmin/user_media_upgrade/Who_we_are/Mission/ISID_SDG_brochure_final.pdf.

United Nations Secretary-General's High-Level Panel on Global Sustainability, 2012. Resilient People, Resilient Planet: A Future Worth Choosing. New York.

WCED, 1987. Report of the World Commission on Environment and Development: Our Common Future (The Brundtland Report). Retrieved October 22, 2015, from. http://www.un-documents.net/our-common-future.pdf.

World Bank, 2017a. Atlas of Sustainable Development Goals 2017. World Development Indicators, Washington, DC. https://doi.org/10.1596/978-1-4648-1080-0.

World Bank, 2017b. Shaping Sustainable Consumption and Production Agenda in Turkey : A Study on Economic Instruments to Support SDG 12.

World Bank, 2018. Atlas of Sustainable Development Goals 2018. World Development Indicators, Washington, DC. https://doi.org/10.1596/978-1-4648-1250-7.

CHAPTER

Smart cities as drivers of a green economy

5

Osman Balaban

Introduction: the double crisis

Today the world is facing a double crisis: the economic crisis and the ecological crisis. Both crises are two separate but highly interrelated forms of disturbances that urge us to rethink the ways we utilize the planet's resources to meet our needs. The prevailing patterns of production, consumption, and circulation in the capitalist economy have to be changed if the future is to be sustainable, inclusive, and prosperous for all.

The economic crisis that the world is facing is not just limited to the global financial crisis of 2008 and its aftershocks but also the crisis associated with deeper processes and the mainstream ideology on which Western capitalism has been founded. Western capitalism has not been functioning well in recent years (Jacobs and Mazzucato, 2016), and this has turned the global economy into a highly unsustainable and a costly economic system. Today's unsustainable economic system is running mostly, if not solely, on the desire of short-term gross domestic product (GDP) growth. Continuous growth of GDP is the major prerequisite for the world economy. Global GDP, for instance, increased from $13.4 trillion to $34.1 trillion (in constant 1995 dollars), indicating a threefold increase, between 1970 and 2000 (Dauvergne, 2008). The continuous growth of GDP necessitated increased levels of production and consumption. The main recommendations of economic orthodoxy for policy-making focused on expanding the bases of production and enabling the societies to consume, on an equal pace, what has been produced. The outcome has been the tremendous increase in consumption levels of the world's population. Private consumption expenditures increased more than fourfold from 1960 to 2000, during when the global population only doubled (Dauvergne, 2008).

On the other hand, the economic system, which is based on continuous GDP growth, has failed to improve social well-being and reduce inequality. The high levels of growth in global output as well as national GDPs in many economies did not result in equal or fair outcomes for individuals and societies. The divergence of average household incomes from overall economic growth has become a feature of most economies (Jacobs and Mazzucato, 2016). In less developed parts of the world, 2.7 billion people survive on less than $2 per day, around 800 million people suffer from chronic malnutrition, and over a billion people do not have access to

Handbook of Green Economics. https://doi.org/10.1016/B978-0-12-816635-2.00005-5

clean water (Dauvergne, 2008). In the United States (US), for instance, the real median household income was more or less the same in 2014 than it had been in 1990, despite the 78% increase in GDP over the same period (Jacobs and Mazzucato, 2016).

The second crisis, which is the ecological one, is in fact the consequence of the economic crisis. The growth-oriented economic system has been accompanied by serious environmental damage including various sorts of pollution and biodiversity loss due to ever-increasing use of natural resources and generation of waste. The living planet index, which assesses the state of biodiversity in the world's ecosystems, was found to be 30% lower in 2010 compared with 1970, and the ecological footprint of mankind almost doubled since 1966 (Kudelas et al., 2018). Presently, humanity uses the equivalent of 1.5 planets to provide the resources consumed and absorb the wastes generated (Odero, 2013). This is expected to increase to the equivalent of 2 planets by 2030 and 2.8 planets by 2050, if current patterns of production and consumption continue (Kudelas et al., 2018). What is more, fossil-based energy consumption and associated greenhouse gas (GHG) emissions are increasing at a high pace. The atmospheric concentration of carbon dioxide (CO_2) has already exceeded 400 parts per million (ppm) and is still rising very rapidly, which will soon bring us to 2°C of global warming, which is the threshold level of dangerous warming. Odero (2013) truly states that "the world is running an ecological deficit," which is not only unsustainable but also requires a new economic system that is greener and smarter.

The double crisis that humanity faces today is the outcome of the long-lasting conflict or, in simpler terms, isolation between economic development and environmental policies. Since the industrial revolution, economic development has been pursued at the expense of natural and environmental resources, and almost no nation is an exception. For a long time, environmental concerns have been considered as barriers toward economic development and thereby are given almost no priority in economic policymaking. However, the urgency of the current double crisis forces us to rethink the understanding and assumptions that shape the relationship between economic development and environmental protection. The isolation between the two has to come to an end for the future of our planet and our societies to be safe, inclusive, and sustainable. This requires a paradigm shift, or in other words, an alternative economic model that will target more than the short-term GDP growth including, for instance, improving human well-being, eradicating poverty and inequality, preserving resources, and reducing the environmental risks that future generations may encounter. The "green economy" concept provides us with a significant opportunity here. The concept has emerged and been developed as a promising alternative approach to economic development.

The green economy: evolution, scope, and controversies of the concept

Given its potential to reconcile economic motivations with environmental considerations, the green economy can be a response to the double crisis that the contemporary societies face today. In recent years, the efforts to use the concept to address the financial and climate change double crisis have been intensified in the international political arena (Pitkanen et al., 2016). As such, the timing of the widespread emergence of the concept underlines this fact. Although the green economy concept first emerged in the study of Pearce et al. (1989) titled "Blueprint for a Green Economy: Submission to the Shadow Cabinet," the revival and widespread use of the concept coincided with the aftermath of the 2008 financial crisis. The global crisis created a perfect environment for the green economy's international rise (Georgeson et al., 2017), as many countries that had experienced a recession and massive loss of jobs prioritized the search for alternative ways or models for economic development (Kudelas et al., 2018).

Soon after the crisis, the international community, including the leading international organizations, scientific community, and environmental groups, provided significant support to turn the concept of green economy into a new opportunity or pathway that can overcome the crisis (see, for instance, UNEP, 2011; OECD, 2011; World Bank, 2012). The United Nations Environment Programme (UNEP), Organisation for Economic Co-operation and Development (OECD), and the World Bank called for a radical transformation of current development practices and transitions toward a green economy (Davies, 2013). The 40th World Economic Forum 2010 in Davos and the UN Conference on Sustainable Development 2012 (also known as the Rio+20) in Rio were held around the theme of the green economy so as to invite the international community to rethink and reorganize the world economy to improve the state of our planet. As Clark (2013) notes, the Rio+20 meeting emphasizes that economies must be made both green and inclusive and that poverty eradication is the world's most pressing challenge that requires targeted efforts to reach the poor and vulnerable by creating jobs and opportunities. The international debate after the 2008 global crisis focused on proposing the green economy as an alternative model that would increase human well-being while reducing environmental challenges and risks. Therefore, the conceptual basis of the green economy rests on the premise that isolation or separation of economic development and environmental policies is artificial (Barbier, 2012).

There are various definitions of the concept of the green economy. Maybe the first comprehensive and thus the seminal definition have been made by the UNEP as follows: green economy is the "one that results in improved well-being and social equity, while significantly reducing environmental risks and ecological scarcities" (UNEP, 2011, p. 2). After examining the green economy definitions of some international organizations, Borel-Saladin and Turok (2013, p. 219) conclude that the green economy has the potential to "transform the current system of unsustainable economic activity into a future with a healthy environment and a more inclusive economy." A more detailed definition is made by Ahmed (2013, p. 46) as the "one in which growth in income and employment is driven by public and private investments that reduce carbon emissions and pollution, enhance energy and resource efficiency, and prevent the loss of biodiversity and ecosystem services." As per all definitions, the green economy concept implies a transition process in terms of a structural shift from a resource-intensive economy to a resource-efficient economy, which primarily deals with protection of the natural environment and progress toward social inclusion and equity. The green economy concept is thus defined as a low-carbon, resource-saving, and socially inclusive model of the economy.

One controversial issue with regard to the definition of green economy arises from the relationship between the concepts of green economy and sustainable development. The green economy concept has been criticized for overlapping with sustainable development or attempting to replace it (Georgeson et al., 2017). Another group of critics argue that green economy is not a new issue or concept but just another way of phrasing sustainable development. However, there are bold statements emphasizing that green economy should not be considered as a substitute for sustainable development (Borel-Saladin and Turok, 2013). The OECD (2011, p. 11), for instance, states that green growth is "a subset of sustainable development but does not replace it," underlining that sustainable development is an overarching goal and green economy is a tool to achieve that goal. The UNEP and the World Bank reports also portray the green economy and green growth concepts as new pathways toward sustainability. There are other views that even boldly distinguish between green economy and sustainable development. Lorek and Spangenberg (2014) argue that green economy is a new terminology for ecological modernization with a particular focus on efficiency and innovation, and in this respect, green economy concept may not support or fulfill the Brundtland sustainability criteria. Likewise, Georgeson et al. (2017) argue that green economy may go further than sustainable development and confront the artificial separation of environment and economy by creating a policy framework to achieve economic progress with lower environmental impacts.

Another controversial issue is the conceptualization of "growth" within the green economy approach. The green economy concept does not deny or exclude the idea of "growth," which is fundamental in the traditional economy, and in this respect, green economy differs significantly from the concept of degrowth (Gainsborough, 2017). The growth idea of green economy is "green growth." That is the growth of GDP, which is subject to green conditions and focuses on green sectors as new growth engines (Kudelas et al., 2018). Schmalensee (2012), with reference to the UNEP's approach to green economy, states that the greening of economies is not necessarily a drag on growth but rather could serve as a new engine of growth that generates decent jobs. Furthermore, the rapid development in emerging economies is indicated as opportunities for green economy transitions (Georgeson et al., 2017).

Accordingly, whether or not green economy would bring about and sustain economic growth is an important question. Zenghelis (2016) is optimistic for maintaining of economic growth under an economic model that is based on decarbonizaiton. He argues that growth will continue as cleaner (non—carbon-based) sources of energy, such as solar and wind, are further utilized and that shift of economies toward knowledge capital and information-based goods and services will also drive growth. Thus, Zenghelis (2016) gives a central and a critical role to innovation in his argument. Perez (2016), like Zenghelis, is also emphasizing the role that technological revolution can play to reconcile sustainability and economic growth by means of a range of innovations to reduce material and energy consumption and increasing the proportion of services and intangibles in GDP. So, to a great extent, green economy is associated with creativity and innovation.

In a nutshell, green economy refers to a model of economic organization that not only reduces GHG emissions and uses resources more efficiently but, while doing these, also generates growth in income levels and job opportunities as well as improves social equity and inclusiveness. This suggests a substantial transition both in the scope of goods and services and in the ways goods and services are produced, distributed, and consumed. Technological revolution stimulated by innovation and creativity is the key opportunity to facilitate the green economy transition.

Green economy transition: application of the concept

Most countries in the world have recently introduced policies to move away from the economic model that regards environmental protection a burden to a green economy, which recognizes ecology as the engine of development (Kudelas et al., 2018). The green economy transition has been mediated by concrete cases and experiments in a variety of sectors (Pitkanen et al., 2016). Whereas some of these experiments remained temporary and incremental in their outcomes (Kemp et al., 2007), some others have turned into key components of substantive transitions transforming practices or even national policies (Geels et al., 2004). To give an idea of the economic value of current green economy initiatives, the revised estimate of global investment in clean energy in 2015 was around $348.5 billion, which is projected to be $7.8 trillion by 2040 (Georgeson et al., 2017).

Table 5.1 presents a compilation of the examples of the green economy transition experiments and actions in different countries at different governance scales (see Kudelas et al., 2018; Pitkanen et al., 2016; Werna, 2012; Albino and Dangelico, 2013). As clearly seen on the table, various initiatives have been launched worldwide to provide support for public, private, and nongovernmental actors to implement green economy in practice. On the other hand, different priorities and understandings of green economy seem to have shaped these initiatives, mainly depending on the specificities of the contexts. The current initiatives indicate the most promising sectors of green economy as transport, land use, building construction, energy, infrastructure, and waste management. Furthermore, innovation is at the core of these initiatives. Research and development sector stands out as an

Table 5.1 A selection of green economy transition experiments worldwide.

Place of initiatives	Level of initiatives	Target sectors	Target actions	Specific targets	
				Quantitative	Qualitative
Mexico	National level	- Building	- Energy-efficient buildings and devices	- Reduce carbon emissions by half by 2050	
United States	National level	- Energy	- Effective use of solar power plants	- Meet 65% of energy and 35% of heat consumption from solar - Reduce harmful emissions by 80% by 2050 - Create 5 million jobs in environmentally friendly technologies	
EU	Regional level	- Energy - Transport - Buildings - Infrastructure	- New green measures in target sectors - Standards for automobile emissions - Subsidies for electric vehicles		
Britain	National level	- Economy	- Adoption of the economy of green technologies as a national development strategy	- Create 100,000 new jobs in green projects	
Germany	National level	- Energy	- System-wide energy transition including extensive use of renewables and nuclear phase out		
	Subnational level	- Agriculture - Land use	- Land use change toward restoration of drained peat lands	- Delivery of cobenefits including CO_2 reduction and other ecosystem services	- Three German states issued Certified Emission Reductions (CERs) on the voluntary carbon market

	Subnational level—Hamburg	- Waste - Energy - Transport - Infrastructure - Buildings	- Various actions and measures in target sectors	- Reduce waste generation and increase recycling - Control per capita water use - Energy-generating and efficient building - 600 renewable energy plants in and around the city - Extensive public transport running on cleaner fuels - Incentives to bike use	- Use of partnership programs in various sectors for effective participation of citizens
France	Subnational level	- Waste	- Organic waste minimization via individual and collective composter sites and computerized monitoring tool	- Rennes Metropole succeeded to become a zero waste conurbation	- Compilation of 10 good practices for cities to minimize waste
	Subnational level—Nantes	- Waste - Energy - Transport - Infrastructure - Buildings - Land use	- Various actions and measures in target sectors	- Encourage water savings and prevent leakage in water supply - Waste recycling and reuse - Energy-generating and efficient building - Renewable energy plants in the city - Multimodal and effective public transport system - Encourage use of bicycles - 100% of residents live within 300 m from a green area	

Continued

Table 5.1 A selection of green economy transition experiments worldwide.—*cont'd*

Place of initiatives	Level of initiatives	Target sectors	Target actions	Specific targets	
				Quantitative	Qualitative
Spain	Subnational level—Vitoria Gasteiz	- Waste - Energy - Transport - Building - Land use	- Various actions and measures in target sectors	- Reduce waste generation and increase recycling and composting - Prevent leakage in water supply - Energy-generating and efficient building - Extensive public transport running on cleaner fuels - Incentives to bike use - Enlarge urban green areas and forestry	- Various activities to raise awareness on environmental issues
Finland	National level	- Building	- Extensive use of wood in building construction - Energy and resource-efficient buildings	- Increase the market share of multistorey wooden buildings from 1% to 10% between 2011 and 2015	
Sweden	Subnational level—Stockholm	- Waste - Energy - Transport - Land use	- Various actions and measures in target sectors	- Reduce waste generation and increase recycling and reuse - 70% of households with district heating powered by renewable energy - Extensive public transport running on cleaner fuels - Incentives to bike use - 40% of city space dedicated to greenery	- Raise awareness of waste management

Australia	National level	- Infrastructure - Labor market	- "Green plumbers" initiative; training of plumbers to give them particular skills on water saving, rainwater harvesting, and use of gray water		- Raise clients' awareness on water savings - Generate new business and job opportunities
Japan	National level	- Technology - R&D	- Expand the market of environmental technologies	- Create 2.2 million jobs in green sectors	- Harmonize advanced technologies, traditions, and social mechanisms with the environment
South Korea	National level	- Industry - Energy - Transport - Infrastructure - Waste - Land use	- Adoption of "green growth" as a national strategy in target sectors		
China	National level	- Energy	- Develop renewable energy generation	- Receive 15% of electricity from renewables by 2020 - Reduce carbon intensity of the economy by 45%	

Sources: Kudelas, D., Domru, E., Stoianov, A., Peters, D., 2018. International experience, principles and conditions for the transition to a "green economy". E3S Web of Conferences, vol. 41, 2018, Article No: 04023, 3rd International Innovative Mining Symposium, https://doi.org/10.1051/e3sconf/20184104023; Pitkänen, K., Antikainen, R., Droste, N., Loiseau, E., Saikku, L., Aissani, L., Hansjürgens, B., Kuikman, P., Leskinen, P., Thomsen, M., 2016. What can be learned from practical cases of green economy? Studies from five European countries. Journal of Cleaner Production, 139 (2016), 666–676; Werna, E., 2012. Green jobs in construction. In: Ofori, G. (Ed.), Contemporary Issues in Construction in Developing Countries. London and New York: SPON Press, 408–441; Albino, V., Dangelico, R.M., 2013. Green cities into practice. In: Simpson, R., Zimmermann, M. (Eds.), The Economy of Green Cities: A World Compendium on the Green Urban Economy, Heidelberg, New York, London: Springer, 99–113.

overarching innovation activity, which would deepen the potentials of the former sectors for green growth.

Cities and the green economy transition

Why cities?

Today, transition to a green economy is widely recognized as an urgent task to give an end to the isolation between economic targets and environmental concerns. Cities are the most appropriate places to this end or, in other words, the most appropriate focal points of green economy transitions. The growing size and importance of urban areas make the city the most important entity for fostering the green economy (Puppim de Oliveira et al., 2013). Cities are now the hot spots of population and economic growth, which makes them the engines of the global economy. As of 2014, 54% of the world's population lived in cities and the share of urban population is expected to rise to 67% by 2050. Besides, major economic activities are located in cities, and this results in generation of the lion's share of the national and global GDPs in urban areas. For instance, the richest 100 urban economic areas are estimated to produce almost 30% of global GDP in 2008 (Simpson, 2013), and more to that, 46% of the world's gross value added in 2007 was generated in the 150 of the world's most significant metropolitan economies that accommodated only 12% of the global population (Berube et al., 2010).

On the other hand, cities are at the same time responsible for most environmental problems as a consequence of their production and consumption patterns. Urban footprints on the globe are increasing day by day. Although cities occupy only 2% of the earth's land surface, they account for around 75% of the global energy and resource consumption as well as the associated CO_2 emissions (Simpson, 2013). As mentioned earlier, the world is running on an ecological deficit, and maybe the most important part of this deficit is created in cities.

Thus, it is, in a sense, an urban task to reconcile the economic processes and environmental conditions. If not cities, where else can we take specific and concrete actions to facilitate the transition toward a greener economy? The concentration of people, knowledge, infrastructure, economic activities, and resources in cities provides unique opportunities to create an economic environment where efficiency is increased significantly and economic outputs are not generated at the expense of environmental problems (Simpson, 2013; McCormick et al., 2013).

The urban transition: smart cities as drivers of a green economy

As obvious, the double crisis that we face today can be overcome by a substantial transition process. This transition will take place along two different but highly interrelated pathways. Along the first transition pathway, today's resource- and waste-intensive economy will be replaced by a green economy that respects the ecological limits and concerns, and responds to the needs of the disadvantaged and the poor. This is the pathway for green economy transition.

The green economy transition requires another transition pathway that targets cities and urban areas. Considering the central role that cities play in organization of the global economy and occurrence of the "ecological deficit," the second transition pathway should target to create the urban environment that stimulates the transition to a green economy. This is the urban transition pathway.

The urban space and the ways urban space is produced and used have to go through a structural transformation to provide the necessary material basis as well as a range of investment fields for the green economy. The role of cities in transition to a green economy involves, on the one hand, the greening of the city-based economic processes and, on the other hand, the greening of the production and use of urban space itself. As cities face different challenges, urban transition to a greener economy cannot follow a uniform pathway. The specific conditions and urgent needs of the local contexts have to be considered. On the other hand, the current initiatives for greening cities and urban economies share a common vision for the future of cities. Odero (2013), for instance, suggests urban planners to shift their attention from "asking where economic activities are taking place, to what kinds of activities and toward what end?" Yet, this way of defining the common vision is insufficient. It should be taken one step further to encompass the shift of the focus of urban planning from determining where urban functions are taking place to strengthening the connections and increasing the synergies between the key constituents of an urban system.

There are cities that have already embarked on the pathways of the transition process to overcome the economic and ecological crises. The existing practices in these cities highlight some specific urban sectors and some particular actions in these sectors (see Table 5.1). Among the most significant of these sectors and actions are mixed and compact land use, renewable energy, low-carbon transport, green buildings, clean technologies in urban service delivery, improved water provision, urban farming, and improved waste management. Technological change based on innovation and creativity is the cross-cutting component of these initiatives and actions.

Considering the specific sectors of green economy transition in cities and the central role that technological progress and innovation can play in this transition, "smart city" stands out as the most appropriate concept that could shape the urban transition to a green economy. In other words, the conceptual foundation of the city of the green economy can be based on the idea of smart city. The smart city idea emerged in the 1990s in parallel to the advancements in computer science and information and communication technologies (ICTs) (Albino et al., 2015).

In general, smart city can be defined as the city where the state-of-the-art ICTs is applied to the design of city space and to the provision of major urban services so as to make the city more efficient, sustainable, and livable. As seen, the emphasis is on the use of the most advanced technologies in design and organization of an urban system. The narrow definitions of the concept focus on hard and technical issues where application of technology is more straightforward. An example of a narrow definition is as follows: "a city of high efficiency with integrated infrastructure by leveraging IT to support urban life" (Fukuchi, 2011 cited in Kono et al., 2016).The broad definitions focus also on soft issues such as governance, people, and living, along with the hard and technical ones. For instance, Giffinger and Gudrun (2010) provide a broader definition by identifying the six essential characteristics of a smart city as a smart economy, smart mobility, a smart environment, smart people, smart living, and smart governance. In parallel to the progress in academic literature, the smart city concept goes beyond the narrow definitions to emphasize that ICTs should not be considered on its own but should be embedded in wider physical and social systems for allowing the technological progress to be at the service of people, business, and government (de Jong et al., 2015).

Integration and connectivity are the key features of smartization of urban systems and communities. Smart city is, in a sense, the city, where almost all key components of the urban system are connected and integrated to each other in strong ways. Enhanced and widespread connections of the system components create synergies and optimize the system performance, which bring about efficiency increases and significant savings. In this context, smart city concept relies on the Internet of things, which is a recently evolving technology platform that connects every device in the planet with any and every other thing or device (Addanki and Venkataraman, 2017). The real-time data generated by multiple sources can be processed accordingly to provide simultaneous, sequential, or cumulative feedback to support communication and interaction among citizens, service providers, and institutions. In this way, the major constituents of urban systems are enabled to make more optimal choices and sustainable decisions. The use of technology to enhance the connection and integration in cities offers concrete innovation and investment opportunities for physical urban and infrastructure development and promotes engineering system solutions to urban problems (de Jong et al., 2015), and thereby provides various channels of green economy transitions.

All in all, we have two important transition processes in front of us. The first one applies to the economy, implying the transition from today's mainstream economic model to a green economy. The second one applies to the production and use of urban space, implying the transition from today's cities to the smart cities of future. The following sections discuss the roles that the four key urban sectors, namely land use, buildings, transportation, and waste management, can play in transitioning to a smart urbanization and a green economy.

Urban land use

In general, land use refers to the use of urban space by inhabitants of cities. In particular terms, it is the physical basis on which major aspects of urban life materialize. In the context of urban planning and development, land use structure of a city defines the organization of all key functions and activities that form and shape an urban system (Balaban, 2017). The mainstream economic system and its associated economic policies have significant impacts on land use structure of the world's cities. Until recently, Western capitalism disseminated the idea that environmental protection and economic growth were incompatible and environmental protection was a barrier to economic success. The assumption that cities, environment, and economic growth are incompatible has shaped the priorities of city administrators in the use and organization of urban space. As cities primarily focused on economic growth, which is meant having a substantial industrial base, top priority in land use planning was usually given to the provision of the infrastructure to service industry (McKendry, 2013). As such, the footprints of most cities built after the World War II are highly unsustainable and resulted in the rise of concerns about city greening during the past several decades (Odero, 2013).

The transition to a greener economy by means of smart cities necessitates a fundamental change in urban land use planning. The city administrations have to reconsider their priorities in the production and use of urban space. The environmental concerns, which were for long seen as a barrier for local economic development, now can be seen as opportunities for green growth. Such an understanding should determine the agenda of developing new urban spaces as well as renewing the existing ones in the transition to a green economy.

From the land use perspective, smart city, as the driver of a green economy, is a compact city that consists of mixed-use quarters, which are strongly connected to each other and to the entire city system via a range of transport and communication links. A compact city provides the city administrations with significant efficiency increases and cost savings mainly due to the increased proximity between urban residents and activities. Compared with sprawled patterns of urban development, provision of public transit systems and integration of various transport modes can be realized in compact cities, which in turn bring about substantial reductions in energy use and carbon emissions from urban mobility. For instance, at similar population levels, there is a 6-fold difference between the transport-related per capita CO_2 emissions in Atlanta (US) and Barcelona (Spain) due to the 12-fold difference between the land area occupied by both cities (Litman, 2015; Balaban, 2017). Mixed-use quarters within the compact city also support such efficiency increases and cost savings. The idea behind mixed-use quarters is to shorten the distances between the working, living, and shopping places of residents. Mixed-use quarters enable their residents to live, work, and satisfy their daily needs in walking and cycling distances. Rode et al. (2014) argue that a reduction of by up to 25% in vehicle-kilometers-travelled can be achieved by doubling densities and also concentrating employment within densified areas in the US metropolitan regions.

Creating urban quarters as innovation clusters

Cities are incubators of innovation, as they benefit from the concentration of diverse but specialized skill sets in research institutions, firms, and service providers that can pilot and scale new technologies (Rode, 2013). New patents, for instance, are granted excessively in larger urban centers (Bettencourt et al., 2007), and as per the OECD data, urban regions in the OECD accounted for 73% of green patents in the renewable energy sector over 2004–06 (Kamal-Chaoui and Robert, 2009).

A highly networked urban environment is crucial for cities to function as agents of innovation and creativity. City administrations should prioritize the development of networking platforms for enhanced knowledge sharing and creation of eco-innovation and green clusters by facilitating research collaboration and synergies (Kamal-Chaoui and Robert, 2009).

Berlin has a good example of such a networked urban quarter, namely the EUREF Campus, where companies from energy, environment, and mobility sectors are settled to work for developing ecologically and economically sustainable solutions. Soon after its launch, the EUREF campus has become an important center for innovation and communication to stimulate energy transition in Germany and also in Europe (Euref, nd.). InfraLab is an interesting and a promising initiative of the EUREF Campus for achieving resource and energy efficiency in Berlin. It is the coinnovation lab that brings together the six major utility companies of Berlin to stimulate the dialogue and cooperation among them for achieving a more sustainable and efficient service provision to Berliners (Infralab, nd.).

Such networked environments in cities create new synergies or deepen the already existing synergies between different constituents of the urban systems. In this way, not only significant efficiency increases in energy and resource use are achieved, but also a circular economy in which outputs of one sector become the inputs of others can be created (Rode, 2013). Therefore, cities should work for creating highly networked urban quarters as clusters of innovation, synergies, and symbioses as part of their agenda for green economy and smart city transitions.

From industry-based solutions to nature-based solutions

The ecological footprints of the industrial city and the associated concerns about city greening have changed our understanding of the relationship between city space and nature. The approach to urban green spaces in land use planning started to shift from allocating patches of arguably green areas to bringing back nature and nature-based systems into the city space. The notion of integration and connection of the smart city concept is also applicable to this new way of green space planning. One of the main constituents of urban land use in a smart city is the network of green areas in which enhanced connection among urban greenery creates continuous channels of natural corridors in and through the city space. The continuity of green spaces provides hot spots of biodiversity preservation and also enhances the ecosystem services that the entire city receives from its green spaces. In case it is not possible to ensure green space continuity and connectivity due to presence of urban built structures, roofs and facades of the buildings can be turned into vegetated spaces. Smart city applications via (thermal) cameras to detect temperature changes or follow the presence and movements of species along green spaces can provide real-time and cumulative data to be used to assess the impacts of the green space network on biodiversity preservation, and ecosystem services.

Thus, the change in the way we look at the production and use of urban lands from a smart and green city perspective promises new business and investment opportunities that would generate new jobs and income while reducing environmental risks and challenges.

Urban buildings

Cities accommodate millions of buildings of different types, functions, and sizes. Buildings are the core components of urban systems not only because they stay with us for long years once they are constructed but also because most aspects of urban life are realized in and through buildings. In this regard, the building sector has a vast impact on the natural environment and the economy (Jiang, 2017). Especially the ecological and carbon footprints of the building sector are remarkably high and still in an increasing trend. The building sector is the largest energy consumer in the European Union (EU) and the United States with a share of 37% in total final energy use in both contexts (Juan et al., 2010; Perez-Lombard et al., 2008). High use of energy in the building sector has also increased the sector's carbon footprint, estimated to be 15% of the global GHGs (Baumert et al., 2005).

The ever-increasing impacts of buildings on the natural environment have made the building sector a major focus of city greening and smartization agendas. The green building concept has been developed as a response to address environmental problems that stem from buildings and reduce the impacts of the building sector on natural environment (Balaban and Puppim de Oliveira, 2017). There are other concepts such as low-carbon buildings, ecobuildings, or sustainable buildings, which are used interchangeably with the concept of green buildings. Although there are slight differences between all these concepts, they jointly refer to application of sustainability principles as well as advanced technologies to design, construction, management, and even demolishment stages of buildings so as to maximize the resource and energy efficiency and minimize the impacts on the natural environment.

In many cities, considerable efforts and resources have been spent to make buildings more energy and resource efficient by means green building construction and retrofitting. Tokyo Metropolitan Government introduced the world's first city-based Cap and Trade Program in 2010 as a mandatory emission reduction system that applied to the 1300 large CO_2 emitting facilities including all of the high-rise buildings in Tokyo (Nishida and Hua, 2011). Establishing of green building certification systems and their associated councils such as the LEED, BREAM, and CASBEE in many countries has supported the growth of the green building industry and the development of green building technologies in the past decades. The rising energy costs of building construction and operation together with the potential cobenefits including substantial cost savings that greening of the buildings can deliver have generated great interest among private and public actors of cities. While some large-scale corporations invested in greening the headquarter buildings and premises that they occupy, several city administrations renewed their building codes and regulations to enable green solutions in the building sector. For instance, the Tokyo Green Building Program requires the owners or developers of all new buildings with total floor space exceeding 5000 square meters to publicize their building's environmental plan and performance evaluation with the aim of encouraging building owners to apply green design principles (Balaban and Puppim de Oliveira, 2017). All these efforts and initiatives have made green buildings one of the major items of city greening agenda. This particular achievement has resulted in significant reductions in energy consumption and CO_2 emissions in buildings.

Greening of urban buildings highly relies on technology development. Performance and efficiency increases in buildings' energy and resource use arise from the smart and innovative solutions applied. A part of these solutions is based on the use of computerized systems that connects all major components of buildings to collect data and identify patterns of energy and resource use by the occupants. The data and patterns are then used to manage the systems that supply energy or resources to the building in more optimal and efficient ways. A good example to such computerized systems is the Building Energy Management Systems, which are standardized energy management systems that use computerized information processes to reduce energy consumption effectively in a building without compromising the comfort and safety of its internal environment (Nakagami, 2011).

Innovation not only is limited to the high technology active design solutions in building sector but can also bring about significant passive design solutions that improve environmental performance of buildings at relatively lower costs than active design solutions. In Japan, for instance, ecovoid is a common passive design solution in new green buildings. Ecovoid is an empty channel in the middle of the building, which, in some cases, is combined with sun-tracking sensors and mirrors on rooftop, designed to bring more natural light and fresh air into the building. An ecovoid system is calculated to deliver an electricity saving of 15% in an example building (Balaban and Puppim de Oliveira, 2017). Innovation for passive design is highly relevant for developing country cities where resources are limited and high-cost technology solutions may be difficult to afford. For example, in Puerto Princesa City in the Philippines, passive design techniques such as increased natural light, improved ventilation, the cooling effect of the roofing material, and strategic planting have been used to reduce energy demand in coastal housing projects (Rode, 2013).

In a nutshell, greening of urban buildings either through new constructions or retrofitting is an important step in greening and smartization of urban systems, and it opens up new business and investment opportunities that are barely available in the conventional economy.

Urban transport

Transportation is an important emitter of GHGs and also source of various environmental problems, particularly a major cause of air pollution. The transport sector is known to be responsible for 21% of world's energy-related CO_2 emissions, and this share is expected to increase up to 23% by 2030 (Grazi and Van den Bergh, 2008). In Delhi, vehicular pollution is the major contributor to air quality problems with 64% contribution to total pollution in the city in 1991 and 70% in 2000–01 (Sidharta,

2003). Therefore, among the most common and significant steps to make cities greener is to improve the transport infrastructure and systems in cities. Many cities across the world have already been investing in concrete policies and projects ranging from rail-based public transport projects to provision of bus lines and cycle paths and to the use of smarter and cleaner vehicles.

The most common green transport strategies in cities focus on reducing car use, thus encouraging the use of public and nonmotorized options. However, the latter modes are not as easy to use as private modes due to general comfort problems and the last mile issue, which refers to the difficulty in getting people from a public transportation stop to their final destination. The difficulty in the use of public and nonmotorized modes is attempted to be overcome by smartization of these systems. Computer applications that indicate the exact time for departure and arrival of vehicles and duration of and the stops along journeys provide the users with ease and time savings. Smartization of transport systems also focuses on multimodal integration so as to connect all transport modes in a locality in strong ways. Although multimodal integration brings about time savings, its main benefits come in terms of comfort improvements and addressing of the last mile issue. Public transit systems are structurally more efficient than private modes due to larger carrying capacities and economies of scale. However, the efficiency of these systems can further be improved by means of technological development and smart solutions. The Delhi Metro is a good example. The metro system in the Indian Capital is equipped with a regenerative braking system on its rolling stock, which generates electricity when brakes are applied on trains in braking mode and feeds it back into the system to be used by trains in other modes of operation, and in this way, almost 35% of electricity consumed in Delhi Metro is regenerated by the system (Doll and Balaban, 2013).

Smartization schemes apply for individual or private modes and make them cleaner and less harmful. Maybe the most important step in this regard is development of electric vehicles (EVs). As long as electricity supply comes for clean and renewable sources and EVs have longer driving distances, private car traffic in streets of cities of future may not be as polluting as that of the cities of the past. Greening of private modes by means of smart solutions is very crucial in the context of aging societies and small communities. Use of public transport becomes very challenging at old ages due to the access and egress difficulties to stations, platforms, and even to the vehicles. Likewise, in small communities and towns, it is not feasible to construct public transit systems. Therefore, the need for private modes will continue in the future, at least among aging societies and small communities. There is a substantial green economy potential at this point. Technology firms and R&D companies have been working for developing autonomous vehicles or smart micromobility options to address mobility issues of senior citizens in small communities. The use of EVs, smart autonomous, or micromobility vehicles requires the city space and transport infrastructure designed or changed in appropriate ways. Road infrastructure and the parking spots need to be designed or renewed in ways to provide fast charging spots or wireless charging systems. More to that, in smart cities of a green economy, car ownership is highly likely to be replaced by car-sharing systems, which will allow urban residents to borrow smart and clean vehicles when necessary. Car-sharing systems have the potential to solve the last mile issue and thereby encourage the use of public transit systems.

Just like the building sector, urban transport sector is another fertile channel for greening the economy, as smartization and greening of urban transport systems create various fields of green investments and jobs.

Urban waste

An obvious consequence of today's unsustainable economic system is generation of high amounts of waste. Urban areas, as centers of production and consumption, are major sources of waste generation worldwide. The municipal waste generation, which is 1.3 billion tons per year, is expected to increase to 2.2 billion tons per year by 2025, indicating an annual rate of 5% and an increase in per capita municipal solid waste from 1.2 to 1.42 kg by 2050 (Dashti, 2017). These facts indicate and increase the role that the waste sector can play in transition to a green economy via smartization of urban systems. Waste management offers significant cobenefits to public and private actors in cities. These cobenefits range from reduction in pollution and GHG emissions to energy and income generation. In the recent decades, the economic potential embedded in the waste sector has enabled city administrations to develop innovative and new methods of waste management.

Many cities have demonstrated considerable achievements in finding green and smart solutions to reduce overall waste through new forms of environmentally friendly treatment of unavoidable waste (Rode, 2013). These achievements also indicate that effective investments in waste sector can help reduce GHG emissions substantially. The green and smart solutions in waste management range from recycling and reuse to composting of waste, from incineration of organic wastes to landfill gas capture. For instance, in many European cities, recycling levels are in the region of 50%, whereas Copenhagen only sends 3% of its waste to landfills (Rode, 2013). Yokohama city in Japan has introduced an effective waste management plan to reduce the amount of wastes generated and landfilled in the city. Since the launch of the plan in 2003, per capita amount of solid wastes generated by households and enterprises were reduced from 465 to 260 kg between 2001 and 2008, despite the increase in the city's population by 170,000 people over the same period (Balaban and Puppim de Oliveira, 2014). More to that, Yokohama city has also managed to reduce 840,000 tons of CO_2 equivalent of GHGs between 2001 and 2007 due to reduction in waste generation and reuse of solid wastes (Suzuki et al., 2010).

There are good practices in the developing world as well. Indonesian cities are quite well known for their considerable efforts for composting solid wastes. In Surabaya city, for instance, effective composting facilities of the local community in collaboration with the local government managed to reduce solid waste that had gone to landfill by 30% in 2010, generating significant cobenefits as well. Cost saving by the municipality was estimated as 4 million US$ per year, and reduction in GHGs from composting was calculated as about 2800 tons of CO_2 equivalent in 2010 (Premakumara, 2013). In a related development, there are a number of municipal waste-to-energy projects in Turkey, mainly stimulated by a financial support mechanism for renewable energy generation introduced by the national government in 2005. Many cities have already upgraded their waste collection systems and landfill sites in appropriate ways to generate electricity from landfill gas. As of 2017, there are 32 waste-to-energy (landfill gas) plants licensed in Turkish cities, and the installed capacity of these plants is about 195 MW (Balaban, 2018).

To sum up, given the current progress in waste management in many parts of the world, waste sector can be considered as the readiest among the four key urban sectors for green economy transition.

Discussion

The current economic system is highly unsustainable. It has created an economic crisis and an ecological crisis, which seriously threaten the future of our planet and societies. Both crises could be overcome by transitioning to a green economy, which respects natural boundaries and ecological limits, and responds to the needs of the poor and the disadvantaged. Green economy transition requires another transition pathway that applies to the production and use of urban space. Given the role that innovation and technology have to play in green economy transition, it seems plausible to refer to the smart city concept for outlining the city of a green economy.

The smart city concept is based on the idea that the state-of-the-art computer technologies and ICTs should be incorporated into the design, management, and use of all relevant components of urban systems. In this way, the major constituents of urban systems are connected to and integrated with each other, which would, in turn, bring about system optimization, efficiency increases and savings in resource and energy use, and reduction in waste generation. Smartization of urban systems takes place by means of a range of activities and experiments in key urban sectors such as land use, buildings, transport, energy, and waste. These activities and experiments bring about various investment fields, income generation, and job creation opportunities in green economy sectors. Thus, smart city idea can help us not only produce the physical basis on which a green economy materializes but also create new investment and business channels to facilitate the transition processes to a green economy.

Nevertheless, smartization of urban systems and transition to a smart city are not problem-free processes. Instead, they entail substantial challenges to consider and address. If not properly managed, smart city development process may deepen the existing inequalities or create new forms of inequalities among the world's cities and social groups located in cities. This is definitely against the social justice and inclusiveness targets of the green economy concept.

Smart city applications are based on the widespread use of computerized devices and applications, which are not easily available at present. Several social groups, even in cities of developing countries, may not afford such devices and applications. Even if affordability issue is solved, many citizens may not be able to use them. Hence, the risk of societal exclusion is high, and it needs to be considered.

So far, private sector is leading the smart city experiments in many cities. The current top-down implementation of the smart city concept has created the hegemony of global technology firms (Viitanen and Kingston, 2014). The obvious motivation of private sector in these experiments is profit seeking. However, not every city promises the same level of profit-making potential to private sector. Cities in less developed parts of the world are still suffering from the basic problems of rapid urbanization and urban migration. Private sector is, naturally, selective in its destinations in smart city transition processes. So, what will happen in cities that do not promise high returns to investments to smart city initiatives is an important question. There is an important risk for deepening the uneven development patterns among nations and cities. That is why the risk of uneven development also deserves substantial attention.

Smart city systems are data driven, and citizens create much of these data through their everyday transactions, mostly without their knowledge or informed consent (Viitanen and Kingston, 2014). Smart city transition is possible with the creation of these big data, which consists of an enormous amount of knowledge and information with regard to every aspect of urban life and living practices of urban residents. More to that, smart city transition process itself also creates big data, which are collected and managed mostly by private sector and partly by public bodies. However, the current initiatives for smart city development usually indicate that big data are not shared with the public and mostly owned by private firms. Big data constitute a potential threat as the data on private aspects of people's life, their choices, and expectations can be used for manipulation. So, the risk of manipulation of people's choices is another crucial problem to deal with.

All in all, the smart city transition pathway has also a flip side, which reveals significant risks and threats. Without considering these risks and addressing them in proper ways, the smart city idea cannot support the transition to a green economy. If not the state, who else can take the lead and play important roles to minimize the risks and threats intrinsic to smart city transition?

References

Addanki, S.C., Venkataraman, H., 2017. Greening the economy: a review of urban sustainability measures for developing new cities. Sustainable Cities and Society 32, 1–8, 2017.

Ahmed, E.H.M., 2013. Green cities: benefits of urban sustainability. In: Simpson, R., Zimmermann, M. (Eds.), The Economy of Green Cities: A World Compendium on the Green Urban Economy. Springer, Heidelberg, New York, London, pp. 45–56.

Albino, V., Dangelico, R.M., 2013. Green cities into practice. In: Simpson, R., Zimmermann, M. (Eds.), The Economy of Green Cities: A World Compendium on the Green Urban Economy. Springer, Heidelberg, New York, London, pp. 99–113.

Albino, V., Berardi, U., Dangelico, R.M., 2015. Smart cities: definitions, dimensions, performance, and initiatives. Journal of Urban Technology 22 (1), 3–21.

Balaban, O., 2017. Land use. In: Doll, C.N.H., Puppim de Oliveira, J.A. (Eds.), Urbanization and Climate Co-benefits: Implementation of Win-Win Interventions in Cities. Routledge, London and New York, pp. 67–87.

Balaban, O., 2018. Local capacity matters: challenges and opportunities to enhance co-benefits of climate policies in Turkish cities. Paper Presented at the Cities & Climate Change Science Conference, March 5–7, 2018, Edmonton, Alberta, Canada.

Balaban, O., Puppim de Oliveira, J.A., 2017. Sustainable buildings for healthier cities: assessing the co-benefits of green buildings in Japan. Journal of Cleaner Production 163, S68–S78.

Balaban, O., Puppim de Oliveira, J.A., 2014. Understanding the links between urban regeneration and climate-friendly urban development: lessons from two case studies in Japan. Local Environment 19 (8), 868–890.

Barbier, E.B., 2012. The green economy post Rio+20. Science 338 (6109), 887–888.

Baumert, K.A., Herzog, T., Pershing, J., 2005. Navigating the Numbers: Greenhouse Gas Data and International Climate Policy. World Resources Institute, Washington, D.C.

Berube, A., Friedhoff, A., Nadeau, C., Rode, A., Paccoud, A., Kandt, J., Just, T., Schemm-Gregory, R., 2010. Global Metro Monitor: The Path to Economic Recovery. Washington, DC and LSE Cities. In: Metropolitan Policy Program. The Brookings Institution. London School of Economics and Political Science, London.

Bettencourt, L.M., Lobo, J., Strumsky, D., 2007. Invention in the city: increasing returns to patenting as a scaling function of metropolitan size. Research Policy 36 (1), 107–120.

Borel-Saladin, J.M., Turok, I.N., 2013. The green economy: incremental change or transformation? Environmental Policy and Governance 23, 209–220.

Clark, H., 2013. What does Rio+20 mean for sustainable development? Development 56 (1), 16–23.

Dashti, M., 2017. Waste. In: Doll, C.N.H., Puppim de Oliveira, J.A. (Eds.), Urbanization and Climate Co-benefits: Implementation of Win-Win Interventions in Cities. Routledge, London and New York, pp. 170–190.

Dauvergne, P., 2008. The Shadows of Consumption: Consequences for the Global Environment. The MIT Press, Cambridge, MA.

Davies, A.R., 2013. Cleantech clusters: transformational assemblages for a just, green economy or just business as usual? Global Environmental Change 23, 1285–1295, 2013.

De Jong, M., Joss, S., Schraven, D., Zhan, C., Weijnena, M., 2015. Sustainable-smart-resilient-low carbon-eco-knowledge cities: making sense of a multitude of concepts promoting sustainable urbanization. Journal of Cleaner Production 109, 25–38, 2015.

Doll, C.N.H., Balaban, O., 2013. A methodology for evaluating environmental co-benefits in the transport sector: application to the Delhi metro. Journal of Cleaner Production 58, 61–73.

Fukuchi, M., 2011. Conditions on Domestic/overseas Smart Cities. Chitekishisan Sozo, 2011 May (In Japanese).

Gainsborough, M., 2017. Transitioning to a green economy? Conflicting visions, critical opportunities and new ways forward. Development and Change 49 (1), 223–237.

Geels, F.W., Elzen, B., Green, K., 2004. General introduction: system innovation and transitions to sustainability. In: Elzen, B., Geels, F.W., Green, K. (Eds.), System Innovation and the Transition to Sustainability: Theory, Evidence and Policy. Edward Elgar, Cheltenham, pp. 1–16.

Georgeson, L., Maslin, M., Poessinouw, M., 2017. The global green economy: a review of concepts, definitions, measurement methodologies and their interactions. Geography and Environment 4 (1), e00036, 1-23.

Giffinger, R., Gudrun, H., 2010. Smart cities ranking: an effective instrument for the positioning of cities? ACE – Architecture, City and Environment 4 (12), 7–25.

Grazi, F., Van den Bergh, J.C.M., 2008. Spatial organization, transport and climate change: comparing instruments of spatial planning and policy. Ecological Economics 67, 630–639.

Jacobs, M., Mazzucato, M., 2016. Rethinking capitalism: an introduction. In: Jacobs, M., Mazzucato, M. (Eds.), Rethinking Capitalism: Economics and Policy for Sustainable and Inclusive Growth. Wiley-Blackwell, The UK, pp. 1–27.

Jiang, P., 2017. Buildings. In: Doll, C.N.H., Puppim de Oliveira, J.A. (Eds.), Urbanization and Climate Co-benefits: Implementation of Win-Win Interventions in Cities. Routledge, London and New York, pp. 102–118.

Juan, Y.K., Gao, P., Wang, J., 2010. A hybrid decision support system for sustainable office building renovation and energy performance improvement. Energy and Buildings 42 (3), 290–297.

Kamal-Chaoui, L., Robert, A., 2009. Competitive Cities and Climate Change. In: OECD Regional Development Working Papers 2009/2. OECD, Public Governance and Territorial Development Directorate, Milan.

Kemp, R., Loorbach, D., Rotmans, J., 2007. Transition management as a model for managing processes of co-evolution towards sustainable development. The International Journal of Sustainable Development and World Ecology 14 (1), 78–91.

Kono, N., Suwa, A., Ahmad, S., 2016. Smart cities in Japan and their application in developing countries. In: Jupesta, J., Wakiyama, T. (Eds.), Low Carbon Urban Infrastructure Investment in Asian Cities. McMillian, Palgrave.

Kudelas, D., Domru, E., Stoianov, A., Peters, D., 2018. International experience, principles and conditions for the transition to a "green economy". In: E3S Web of Conferences, vol. 41, 2018, Article No: 04023, 3rd International Innovative Mining Symposium. https://doi.org/10.1051/e3sconf/20184104023.

Litman, T., 2015. Analysis of Public Policies that Unintentionally Encourage and Subsidize Urban Sprawl. Victoria Transport Policy Institute. Supporting paper commissioned by LSE Cities at the London School of Economics and Political Science, on behalf of the Global Commission on the Economy and Climate. www.newclimateeconomy.net. for the New Climate Economy Cities Program.

Lorek, S., Spangenberg, J.H., 2014. Sustainable consumption within a sustainable economy – beyond green growth and green economies. Journal of Cleaner Production 63, 33–44, 2014.

McCormick, K., Anderberg, S., Neij, L., 2013. Sustainable urban transformation and the green urban economy. In: Simpson, R., Zimmermann, M. (Eds.), The Economy of Green Cities: A World Compendium on the Green Urban Economy. Springer, Heidelberg, New York, London, pp. 33–43.

McKendry, C., 2013. Environmental discourse and economic growth in the greening of post-industrial cities. In: Simpson, R., Zimmermann, M. (Eds.), The Economy of Green Cities: A World Compendium on the Green Urban Economy. Springer, Heidelberg, New York, London, pp. 23–31.

Nakagami, H., 2011. Reducing greenhouse gas emissions in commercial buildings. In: Onishi, T., Kobayashi, H. (Eds.), Low-Carbon Cities: The Future of Urban Planning. Gakugei Shuppan-Sha, Kyoto, Japan, pp. 59–83.

Nishida, Y., Hua, Y., 2011. Motivating stakeholders to deliver change: Tokyo's cap and trade program. Building Research and Information 39 (5), 518–533.

Odero, K.K., 2013. New urban spaces: the emergence of green economies. In: Simpson, R., Zimmermann, M. (Eds.), The Economy of Green Cities: A World Compendium on the Green Urban Economy. Springer, Heidelberg, New York, London, pp. 17–22.

OECD, 2011. Towards Green Growth. OECD Publishing. https://doi.org/10.1787/9789264111318-en.

Pearce, D.W., Markandya, A., Barbier, E.B., 1989. Blueprint for a Green Economy. Earthscan, London.

Perez, C., 2016. Capitalism, technology and a green global golden age: the role of history in helping to shape the future. In: Jacobs, M., Mazzucato, M. (Eds.), Rethinking Capitalism: Economics and Policy for Sustainable and Inclusive Growth. Wiley-Blackwell, The UK, pp. 191–217.

Perez-Lombard, L., Ortiz, J., Pout, C., 2008. A review on buildings energy consumption information. Energy and Buildings 40 (3), 394–398.

Pitkänen, K., Antikainen, R., Droste, N., Loiseau, E., Saikku, L., Aissani, L., Hansjürgens, B., Kuikman, P., Leskinen, P., Thomsen, M., 2016. What can be learned from practical cases of green economy? Studies from five European countries. Journal of Cleaner Production 139, 666–676, 2016.

Premakumara, D.G.J., 2013. Decentralized composting in Asian cities: lessons learned and future potential in meeting the green urban economy. In: Simpson, R., Zimmermann, M. (Eds.), The Economy of Green Cities: A World Compendium on the Green Urban Economy. Springer, Heidelberg, New York, London, pp. 323–335.

Puppim de Oliveira, J.A., Doll, C.N.H., Balaban, O., Jiang, P., Dreyfus, M., Suwa, A., Moreno-Peñaranda, R., Dirgahayani, P., 2013. Green economy and governance in cities: assessing good governance in key urban economic processes. Journal of Cleaner Production 58, 138–152, 2013.

Rode, P., 2013. Cities and the green economy. In: Simpson, R., Zimmermann, M. (Eds.), The Economy of Green Cities: A World Compendium on the Green Urban Economy. Springer, Heidelberg, New York, London, pp. 79–98.

Rode, P., Floater, G., Thomopoulos, N., Docherty, J., Schwinger, P., Mahendra, A., Fang, W., 2014. Accessibility in Cities: Transport and Urban Form. NCE Cities Paper 03. LSE Cities. London School of Economics and Political Science.

Schmalensee, R., 2012. From 'green growth' to sound policies: an overview. Energy Economics 34, S2–S6, 2012.

Sidharta, P.G., 2003. Present scenario of air quality in Delhi: a case study of CNG implementation. Atmospheric Environment 37, 5423–5431.

Simpson, R., 2013. Introduction: a green economy for green cities. In: Simpson, R., Zimmermann, M. (Eds.), The Economy of Green Cities: A World Compendium on the Green Urban Economy. Springer, Heidelberg, New York, London, pp. 13–16.

Suzuki, H., Dastur, A., Moffatt, S., Yabuki, N., Maruyama, H., 2010. Eco2 Cities: Ecological Cities as Economic Cities. The World Bank, Washington DC.

The World Bank, 2012. Inclusive Green Growth: The Pathway to Sustainable Development. Washington D.C. https://doi.org/10.1596/978-0-8213-9551-6.

UNEP, 2011. Towards a Green Economy: Pathways to Sustainable Development and Poverty Eradication – A Synthesis for Policy Makers. United Nations Environment Programme, Nairobi.

Viitanen, J., Kingston, R., 2014. Smart cities and green growth: outsourcing democratic and environmental resilience to the global technology sector. Environment and Planning 46, 803–819.

Werna, E., 2012. Green jobs in construction. In: Ofori, G. (Ed.), Contemporary Issues in Construction in Developing Countries. SPON Press, London and New York, pp. 408–441.

Zenghelis, D., 2016. Decarbonisation: innovation and the economics of climate change. In: Jacobs, M., Mazzucato, M. (Eds.), Rethinking Capitalism: Economics and Policy for Sustainable and Inclusive Growth. Wiley-Blackwell, The UK, pp. 172–190.

Websites & online references

Euref (nd.) EUREF-Campus, Berlin, Germany. https://euref.de/ [last access 25.02.2019].

Infralab (nd.) InfraLab, Berlin, Germany. https://infralab.berlin/ [last access 25.02.2019].

CHAPTER

Environmental justice, climate justice, and the green economy

6

Begüm Özkaynak

Introduction

While the green economy was the organizing theme of the United Nations Rio+20 Conference, it is a term that generated many interpretations over the past decade much like its precursor—sustainable development—which itself was translated into a global worldview first with the Brundtland Report in 1987 and then the Rio Conference in 1992. This is no surprise, given that both concepts are vague, and the ways in which social and environmental objectives are to be achieved often remain unspecified or rhetorical in official documents (Okereke and Ehresman 2015). In fact, beginning with *The Limits to Growth*, the seminal report by Meadows et al. (1972), up to the concept of ecosystem boundaries proposed by Rockström et al. (2009) and beyond, sustainability discussions have always incorporated competing and conflicting views on what is to be sustained, by whom and for whom (Agyeman, 2008; Davidson, 2006; Özkaynak et al., 2004; Redclift, 2018).

The evolution of sustainability thinking across different traditions was recently compiled by both Scoones (2016) and Redclift (2018), who highlight this long-standing tension in the politics of sustainability. Accordingly, sustainability has multiple versions: some focus exclusively on environmental change; others take the more inclusive sustainable development stance of Brundtland; some others hold that technological innovation can ensure a growing economy with reduced ecological impacts; and yet others claim a sustainable economy is necessarily a nongrowing, even shrinking one. So which version of sustainability, and in a related vein, which direction of transformation do we choose? After four decades of discussion, there is almost no consensus on an answer, for it mainly all comes down to politics. Green economy entered the scene as a new notion on top of this debate, at a time when the need for concrete policymaking processes that deal with a triple—ecological, social, and economic—crisis was urgent. Yet as Agyeman (2008, p. 178) noted for sustainability, green economy is also "*at least* about politics, injustice, and inequality as it is about science, technology, or the environment."

Today, there is not much time left for confusion or rhetoric. It is important to be clear and explicit about how a green economy might look and how it can be achieved, primarily because indicators for climate change and biodiversity loss

Handbook of Green Economics. https://doi.org/10.1016/B978-0-12-816635-2.00006-7

are continuously worsening, meaning that we are globally moving away from a green economy (IPCC, 2018). Moreover, local socio-environmental struggles are intensifying, and there is an urgent need to understand local people's concerns in their own socio-political context. People all over the world, organized in groups and networks, refuse to allow the destruction and contamination of their land, water, soil, and air and fight for the kind of world they want to create—the so-called environmental justice struggles (Martinez-Alier et al., 2016).

This chapter posits that merging the green economy with the ideal of environmental justice is instrumental. As Davies and Mullin (2011, p. 794) noted, justice has so far been put aside, because green economy discussions have mostly been "predominantly technical and financial rather than social and political." Why is it important to emphasize justice in conceptualizing a green economy? First of all, justice is an important element in green transformation from a social perspective, since environmental governance will often create winners and losers (Boyce, 2004). Second, environmental justice needs to be at the heart of green economy discussions because differences in the way justice is embedded in greener economy initiatives—based on different empirical and theoretical approaches—would impact how green and just the economy is in day-to-day practices. For instance, environmental problems regarded simply as "technical matters" by corporate or government experts often mean a lot more to many others. Local communities often feel overburdened by environmental damage and left out of decision-making processes, whereas companies believe they did their best by using state-of-the-art technology and compensating locals in monetary terms for impacts. Third, while environmental issues are often depoliticized (Swyngedouw, 2011; Kenis, 2018), environmental justice offers an encompassing and political perspective for operationalizing the transition to a greener economy, for it easily combines the concerns in the justice literature with insights drawn from ecological economics, political ecology, and social movements. A discussion on environmental justice provides strategic opportunities to academics and policymakers in setting goals and pathways consistent with a green transformative agenda that are both transparent and go beyond rhetoric at the local and global levels.

The chapter is organized in six sections. Following this introduction, Environmental justice: origins and evolution section briefly overviews the origins and evolution of the environmental justice concept. Framing environmental justice section introduces an analytical framework that helps thinking about the range of issues pertinent to environmental justice and as such clarifies why environmental justice analysis has a pivotal role in green economy conceptualizations. Major issues of debate and challenges for a green economy section then highlights some major issues of debate in the environmental justice scholarship and discusses challenges for a green economy. Insights from environmental and climate justice struggles are introduced in Insights from the environmental and climate justice movements section to open up a conversation platform between the proponents and critics of the dominant conceptualization of the green economy. The chapter then concludes with some remarks for future research and policymaking.

Environmental justice: origins and evolution

Environmental justice seems in large part to be a quest for justice (what is the right thing to do?) in the environmental domain, but it has deep and multiple roots of its own, and hence is more complex and challenging to address than initially appears. On the one hand, the term environmental (in)justice has strong empirical foundations, grounded in the grassroots rather than the academia. It originally emerged from the civil rights movement in the United States in the early 1980s against the disproportionate dumping of toxic and hazardous waste in low-income areas populated mostly by people of color—mainly due to the unequal enforcement of environmental protection laws in these regions (Bullard, 1994). Activists of local environmental struggles then adopted and used the justice discourse extensively as a constellation of their claims and contestations in the public sphere (Pellow, 2001; Agyeman and Evans 2004; Pulido, 2017). On the other hand, environmental justice as a concept in its abstract form has also become a policy principle to be adopted at the government level (Fraser and Hanneth, 2003; Agyeman, 2008). Here, effort is spent to define the concept in an environmental/public policy setting, by referencing various universal principles that have been shaped by debates around theories of justice, social choice, and moral philosophy (Sikor, 2016). Baxi (2016, p. 11) refers to these two distinct yet somewhat interrelated environmental justice traditions as a matter of contrast between the "experience of injustice"—or a "sense of injustice" proposed by Cahn (1949)—and "theories about justice." A key difference here is that the empirical approach is grounded not in a particular theoretical position or framing of justice, but in actual and concrete claims made by local communities for environmental justice—precisely the point where the concept originated (Fraser, 2009).

Movement-wise, the environmental justice discourse initially focused on the (un) fair distribution of environmental benefits and burdens and rapidly expanded into an organizing principle used by mobilizations that incorporated demands for recognizing rights, and respecting cultural diversity and participation. In this context, Schlosberg and Collins (2014, p. 361) indicate that the "[e]nvironmental justice movement has never been about equity alone; environmental justice has always focused on how injustice is constructed—why those already exposed to other forms of disadvantage are also subject to environmental bads." Through the years, the movement not only redefined the meaning of justice but also reconsidered and shaped what environment means: what initially simply referred to "wilderness" later implied "where we live, work, and play" and then expanded to include "access to ecosystem services and resources" (Agyeman, 2008).

In line with this broader definition, the themes covered by the environmental justice movement also expanded through the years. The justice discourse was applied to a wide set of issues, such as water, energy, food, transport, land use, and climate; in multiple geographies outside the United States; and at various scales beyond the local, at the regional, national, and global levels (Schlosberg, 2007, 2013; Walker and Bulkeley, 2006). The impetus for such shifts in understanding arose again from the people, as communities all around the world kept contesting practices, policies, and conditions that they judged to be unjust (Bullard, 1999; Schlosberg, 2013). Experiencing injustices on the ground also brought a capabilities-based perspective to justice. Broadly speaking, the capabilities-based perspective refers to the demands of individuals and communities against the deterioration of ecosystems and hence the undermining of their basic needs, capabilities, and functionings (Sen, 2009). According to Schlosberg (2007), this latest perspective is key in addressing the connection between human needs and environmental needs, as it also pays attention to the requirements of ecosystems themselves, and not only to those who depend on them.

Theory-wise, the meaning of justice is also plural, as there are many theories of justice, particularly in the Western tradition. Any conceptual interpretation of environmental justice in terms of policy is unavoidably accompanied by the standard ethical and political inquiries of "who gets what, when, why, and how much" (Bullard, 1999, p. 35). The literature in this context is often directed either at the realm of distributive social justice as found in philosophy and theology or at procedural justice as found in the study and application of the law (Kuehn, 2000). Martin et al. (2014, p. 168), for instance, offer a list of canonical figures whose works can be seen as the main grounds on which judgments about justice are based: "Plato's (1974) attempt to articulate a virtuous life, Bentham's (2009) utilitarian principle of achieving the greatest happiness for the greatest number, Kant's (1998) argument that individual rights must come prior to identification of utility, and categorical imperative that we should act in ways that we would be content to become universal laws of action, or Rawls' (1971) egalitarian principles of liberty and pro-poorness." Nussbaum (2007) and Sen (2009) and have also been building on their previous theoretical developments and put forth the capabilities approach as another principle of justice. Accordingly, capabilities and functionings are important for the freedom people actually have to choose from possible lives.

Given these two distinct paths to environmental justice—the empirical versus the theoretical—which one should be taken? Some scholars prefer the empirical tradition as it inherently recognizes the plurality of justice, and "more often than not, stakeholders do not agree on a single definition of what is morally right" (Sikor et al., 2014, p. 525). The empirical approach is also seen as a good starting point since the injustices experienced by communities are usually place specific and contextual (Martin et al., 2013). Consequently, it offers a good sense of what people consider just and unjust in particular governance settings by providing information on the legitimacy of specific elements in environmental policy. Monetary compensation schemes, for instance, that tend to correct an unjust outcome in terms of (re) distribution are not always considered legitimate by local communities who think it would be more just to simply avoid the damage in the first place (Whiteman, 2009). In this context, Sikor et al. (2014, p. 525), referring to Fraser's (2009) analysis, indicate that the empirical approach "investigates how certain notions of justice find support in public discourse, how they may become dominant, and how they may lose support again." Therefore, the identification of injustice on the ground not only is important on its own but also might be central to a theory of environmental justice (Schlosberg, 2007).

If this is the case, however, Sen (2009) then asks, is there need to go beyond the empirical approach? Do we need to have a theory of justice at all for green governance? Sen's (2009, p. viii) answer is worth noting: "to understand the world is never a matter of simply recording our immediate perceptions. Understanding inescapably involves reasoning. We have to 'read' what we feel and seem to see, and ask what those perceptions indicate and how we may take them into account without being overwhelmed by them."

In other words, both approaches to environmental justice are valuable; they overlap here and there, have certain points in common, and contribute to one another. The problem arises in the policy domain, when there are contested visions of what constitutes just. This occurs particularly if dominant notions of justice (e.g., in government policy) are not in line with those of local communities. The various incentive mechanisms used in such contexts have set in motion a vigorous debate in the literature on how to reconcile justice. Cash payments for tree planting, for instance, may be seen unjust by some because they undermine customary practices of environmental stewardship and intergenerational responsibility and just by others as they contribute to poverty alleviation (Sikor, 2013).

The question that inevitably follows is, what should be the guiding moral principles of green economy governance? The discussion above suggests a need to be more careful in differentiating between the frames of justice adopted in the environmental domain. Which frames of justice can be used and which ones should be avoided? Or put differently, how does a green economy speak to the concerns of the environmental justice movement? These questions and a number of closely related concerns will be addressed in the following sections. The aim here is not to come up with definite answers; rather, it is hoped that a critical review of the relevant literature will create an awareness of how shifts in justice frames and implementation can shape policy outcomes for a green economy. The next section presents an analytical framework based on Sikor (2013) and Sikor et al. (2014) for characterizing the key elements of justice (at the risk of oversimplification) that will be useful to compare and contrast alternative justice framings and their rather different implications for a green economy.

Framing environmental justice

A simple conceptual framework that helps thinking about the range of issues pertinent to environmental justice is provided by Sikor (2013) and Sikor et al. (2014), where three key analytical categories are said to be at the heart of environmental justice analysis: the dimensions of justice, the moral notions/principles of justice, and the subjects of justice. While the authors use this three-tier framework to analyze justice in ecosystem governance and conservation, it is also useful for the purposes of this chapter as a first step toward understanding the plural ways in which environmental justice can be understood for green economy conceptualizations. Each category is briefly introduced below.

Dimensions of justice

A vital component of environmental justice analysis involves the concerns inherent to it. As previously mentioned, environmental justice necessarily embraces distributive conceptions of social justice (how equitable is the distribution of environmental risks and burdens). Yet in the environmental realm, exclusive preoccupation with the question of equitable distribution is inadequate in terms of justice, since at least three other concerns are at stake: recognition (what kinds of values and visions matter; how much are the interests and points of view of particular social groups acceptable within dominant discourse), participation (who takes part in the decision-making process and how), and capabilities (how equal are said capabilities) (Schlosberg, 2007).

However, more often than not, certain dimensions are articulated more by particular groups in particular contexts in search of justice. In policy circles, for instance, practitioners do not take justice beyond the traditional territory of distributive justice; in many instances, the demands by environmental movements for recognition and participation go unnoticed (Martin et al., 2016). Sikor et al. (2014) noted that in ecosystem governance and conservation, local and indigenous people usually emphasize issues of recognition and rights, whereas state agencies generally underline distributional issues such as benefit-sharing and compensation. The situation then becomes a conflicted one, based on whose justice dimension counts. In many instances, looking beyond a distributive model of justice, and focusing also on recognition, makes it possible to uncover the structural, institutional, and psychological forms of domination that often define who benefits from the conservation policy. The conflict over fire management in Canaima National Park between the Pemon indigenous people and the Forest Department in Venezuela is a case in point (Rodriguez et al., 2013), showing that policies considerate of indigenous rights necessarily require more reflection on daily practices, types of knowledge, and respect for alternative ways of connecting with nature. Fraser and Honneth (2003) argue that maldistribution is closely linked to misrecognition, which is a cultural and institutionalized form of injustice. Yet on this last account, they also underlines the dangers of decoupling dimensions from one another and rather fosters an integrated perspective: "maldistribution is entwined with misrecognition but cannot be reduced to the latter" (Fraser and Honneth, 2003, p. 3).

Subjects of justice

The second category of the framework relates to the range, and kinds of stakeholders considered the subjects of environmental justice. As posed by Sikor et al. (2014) and Martin et al. (2016), the questions here are who do we care about morally? Do we only consider the harms caused to the current generation? Or also to future generations? To animals and nature? To ecosystems? Finally, is justice about individuals or is there also a community-level justice? Answers to these questions depend on the extent of justice considerations, and while the environmental justice discourse is strongly linked to the principles of social justice, it has some unique features in this regard. According to Lele (2013), environmental thinking contributes to the standard justice discourse in two distinct ways: first, by broadening the idea of justice to include both intragenerational (social justice) and intergenerational justice (fairness to future generations); and second, by establishing the link between environmental justice within human society and justice for nonhuman nature, the so-called ecological or interspecies justice (fairness to other living organisms).

As noted by Sikor et al. (2014, p. 526), "deep green positions … tend to consider nature or nonhuman species as subjects of justice and accord them equal status to human beings (e.g., Naess, 1990). Yet advocates of intergenerational justice put current and future generations side by side, often assigning an egalitarian distribution of rights and responsibilities between them for sustainability (e.g., The Brundtland Report 1987). Seeing the dangers of dichotomizing things as such, Dobson (1998) attempts to cross the boundaries between intra- and intergenerational and interspecies justice by drawing attention to the interdependencies between social justice and environmental sustainability. In a similar vein, while demands for universal human rights typically emphasize rights of individuals, Schlosberg (2013) noted that in some cases it might also be important to recognize the link between justice for individuals and justice for communities.

Moral principles of justice

The third category of the framework asks what moral principles are applied (consciously or not) when making decisions. Political philosophers tend to develop universal principles on the basis of certain social norms, and then explore their applicability to concrete situations and problems. The literature on theories of justice is vast and has taught us a great deal about the crucial role that principles of justice play in governance. Although it is beyond the scope of this chapter to review them all, some relevant principles and potential tensions are briefly touched upon below to facilitate the discussion on a green economy (for a more detailed account, see, e.g., Okereke, 2006).

To start with a general tenet, Rawls (1971) asserts that principles of justice should be decided from an imaginative state behind a veil of ignorance, so that people cannot tailor principles to their own advantage. The application of rules decided in this manner would be fair and hence lead to just distributive outcomes. Rawls describes two other rights-based principles of justice: first, citizens are conceptualized as being free and equal—to the extent compatible with the provision of a similar liberty for others, and second, inequalities—social or economic—are to be permitted only if the worst off will be better off than otherwise. Then, grounded on this conception of people as being equal and free, Rawls delineates the primary goods essential to develop and pursue specific conceptions of a good life. Armed with an equal rights stance, Bullard (1994), the father of environmental justice in the United States, indicates that people are entitled to equal environmental quality regardless of race, color, or national origin—the right to live and work and play in a clean environment.

There might, of course, be other principles that a society in pursuit of justice might follow, such as utilitarianism or egalitarianism as listed by Okereke (2006). From a utilitarian perspective, for instance, "justice is about designing political institutions and the rule-structure of such institutions to meet the greatest possible amount of happiness for the greatest number of people (Mill, 1967; Bentham, 1970; Hare, 1984)" (Okereke, 2006, p. 728). In this context, actions are just if the generated welfare gains outweigh any related losses, even if some stakeholders end up being worse off. Those who find this notion morally problematic might endorse another idea of justice, one based perhaps on egalitarian liberalism, where the provision of equitable resources is secured so that people can achieve their perception of what is good (Dworkin, 2002). An alternative view conceives justice as meeting basic needs on a fair basis, where the goal is to meet fundamental human needs rather than simply distributing and respecting rights. Without undermining the importance of basic human needs, Sen (2009) puts forward freedom to act in accordance with one's own values and objectives as an indispensable justice foundation (see also Ballet et al., 2007).

In a way, this discussion suggests that justice will always be plural, with no single set of principles that will generate consensus in all contexts. Naturally, several notions of justice may be used simultaneously depending on whom they address and in which setting they operate. Yet as Martin et al. (2014) and Sikor and Newell (2014) noted, if particular framings dominate, then this will imply specific understandings of environmental justice, and hence determine the means by which it will be secured. In green economy and environmental governance, the issue might then boil down to "whose environmental justice counts?" There is a need, therefore, to carefully examine the ways in which policies that aim to promote a greener economy define injustices and deal with them. These points emerge as the main issue for discussion in the following section.

Major issues of debate and challenges for a green economy

Agyeman et al. (2002, p. 78) were the first to merge the differing ideas about environment and justice into a conception of "just sustainability," arguing that "sustainability cannot be simply a 'green' or 'environmental' concern, important though 'environmental' aspects of sustainability are. A truly sustainable society is one where wider questions of social needs and welfare, and economic opportunity are integrally related to environmental limits imposed by supporting ecosystems." While the notion is a useful meeting ground for diverse actors in search for a solution to a world in triple crisis, concerns and tensions about how this might be achieved still remain.

Today, virtually all scholars working on sustainability underline the urgent need to act, so the focus in academic literature and official policy documents is shifting from assessing problems to assessing solutions. Yet as previously mentioned, different justice frames might lead to different recommendations for a green economy, and consequently, there is disagreement on potential solutions, especially in relation to the degree and type of interventions needed for a sustainable future. The disagreements are partly theory-driven, as proposals are obviously rooted in wholly different intellectual and disciplinary traditions, and partly originate from the fact that environmental problems are typically complex; they involve various interested parties; and the way justice is considered, if at all, does not resonate well with long-standing concerns and priorities of the environmental justice movement. This is what Walker and Bulkeley (2006) describe as tensions between the universal and particular notions of environmental justice.

According to Martin et al. (2016, p. 254), "[t]hese concerns include questions about how we distribute costs, benefits, rights and responsibilities, questions about how we give voice to different cultures and beliefs, and questions about how we make trade-offs between current and future people, between individual rights and the greater good, and between humans and non-humans." These points are further developed below. Some key issues of the debate where scholarship on green economy and environmental justice intersect are provided under four headings—rights, responsibilities, and liabilities; compensation and rectification; perceptions of risk; and the problem of time and scale—each directing our attention to a number of key nuances in how justice might be delivered.

Rights, responsibilities, and liabilities

Historically, claims by grassroots environmental movements were always multiple and strongly linked to rights issues—human rights, issues of sovereignty, and cultural survival, racial, and social justice including public health and safety (Bullard, 1994). Indeed, both Bullard (1994) and Schlosberg and Collins (2014) noted that concerns of the environmental justice movement and justice dimensions were already documented in the original set of 17 Principles of Environmental Justice adopted at the First National People of Color Environmental Leadership Summit held in Washington, DC, in 1991. The overall statement was clear; in sum, demanding equal protection from contamination and risks (distribution), calling for policies based on mutual respect (recognition), the right to participate (participation), and self-determination (capabilities). Faber (1998, p. 14) added to these points, noting that "the struggle for environmental justice is not just about distributing risks equally but about preventing them from being produced in the first place."

Regrettably, the conceptual underpinnings of current key global environmental regimes are far from respecting these principles. Okereke (2006, p. 254) noted: "although the texts of global environmental agreements accommodate concepts that express egalitarian notions of justice, core policies remain firmly rooted in market-based neoliberal interpretations of justice, which mainly serve to sustain the status quo." Accordingly, the primary focus of today's environmental governance is generally limited to distributional concerns and tends to be market and private sector oriented, based on values of individual liberty and property rights because Rawls' rights-based approach has been simply interpreted as a right to equal opportunity. It is also mostly ex-post, with little focus on preventive action. According to Lucier and Gareau (2015, p. 495), this has also been observed in the Basel Convention since hazardous wastes have been treated as economic "resources" and the toxic wastes trade reframed "as essential for sustainable economic development rather than as a manifestation of global environmental injustice." In contrast, the transition to a green economy that respects environmental justice requires, above all, a clear understanding of rights and the dynamics behind their formation so that the various forms of inequality—unequal economic and political power in particular—can be dealt with (Aasen and Vatn, 2018).

Two other concepts that are important in green economy conceptualizations are responsibilities and environmental liabilities, since the extent to which they are adopted would make a difference in green economy outcomes. In its broadest sense, a moral responsibility for green economy practitioners would be to keep the economy within safe bounds—not only in environmental but also in social terms. This entails an economy that is both just and within planetary boundaries (Rockström et al., 2009), and requires substantial changes in current institutions (Aasen and Vatn, 2018). Then, as noted by Zografos et al. (2014), when there are risk concerns and/or pollution complaints, one way to deal with them would be to claim liabilities from those who produced them in the first place—both a moral and legal obligation in terms of either preventing damage before it occurs (ex-ante) or generating remedies for restoration and reparation after causing it (ex-post), the former being preferred to the latter. In both cases, recognition of liabilities as such would be crucial in a green economy setting, as it would help guarantee that environmental problems are treated as systemic effects of production and consumption and not just accidental market failures. Green economy interpretations might differ on this matter.

The concept of environmental liability is also relevant at the international scale, since "practices of production, trade, and regulation at one site increasingly connect with seemingly distant sites elsewhere through extended supply chains, technology diffusion, and the internationalization of production" (Sikor and Newell, 2014, p. 151). Within this context, the theory of unequal ecological exchange (Hornborg and Martinez-Alier, 2016) holds that economically wealthy and powerful centers of the world economy (the North) operate in ways that enable them to sustain their own high consumption levels while shifting the ecological burden onto less powerful places (the South). According to Kapp (1970), rather than an unintentional market failure, this is a cost-shifting success—an underlying source of environmental conflict and injustice in our time. Indeed, put more broadly, Kapp would see the capitalist industrial system itself as a system of unacknowledged and unpaid cost shifting to poor people today, to future generations, and to other species—the so-called "ecological debt" (from the North to the South). Although not often recognized in mainstream green economy discussions, referencing ecological debt has the potential of enriching sustainability transformation discussions by provoking a paradigmatic shift if adopted. Therefore, it is important to understand what alternative green economy conceptualizations have to say, if anything, about ecological debt.

Compensation and rectification

If we accept that humankind is ethically obligated to current and future generations, and there are rights, responsibilities, and liabilities assigned to respective agents, the problem then becomes one of specifying a fair policy that will undertake all these endeavors. How might policies aimed at promoting a greener economy ensures that rights and liabilities are taken into account and justice restored? O'Neill (2017, p. 1044) noted that so far "[monetary valuation] has dominated policy responses to environmental problems from climate change to biodiversity loss," whereas there are huge disagreements over its use as a policy instrument to address injustice. Monetary compensation and how it relates to environmental justice are also at the heart of academic and public policy disputes, partly distinguishing the spectrum of justice perspectives that alternative green economies entail (Stevenson, 2015).

O'Neill (2017) differentiates between two legal contexts of monetary compensation: namely, forward-looking cases in policymaking and backward-looking cases of rectification for environmental harms done in the past. In general, compensation (monetary or other) for past environmental harms is normally understood as righting or rectifying a wrong that was committed, and failures to do so are regarded as a major injustice. Yet, offers of monetary compensation for future activities that will harm the environment are the subject of protests by those affected, who claim that prospective wrongs are not open to monetary compensation. In general, mainstream economic approaches calculate potential environmental damage via economic valuation techniques, mainly by assigning a monetary value to aspects and functions of the ecosystem. Instead of identifying the monetary value of the losses due to actual damage (ex-post), extended cost-benefit analyses are used ex-ante to assess whether the damage is worth it. In welfare economics, this is known as the Kaldor-Hicks compensation test (O'Neill, 2017). The environmental justice movement challenges this dominant paradigm. While highly accepted in academic circles on the basis of fears that policymakers do not pay attention on noneconomic arguments, the practical limitations of thinking in terms of cost-benefit analyses and economic valuation in efforts to properly weigh liabilities in environmental justice considerations are well documented (Gerber et al., 2014; Zografos et al., 2014).

The words, as reported by O'Neill (2017, p. 1044), of an indigenous who was offered compensation for displacement due to the Sardar Sarovar Dam in western India are self-explanatory: "You tell us to take compensation. What is the state compensating us for? For our land, for our fields, for the trees along our fields. But we don't live only by this. Are you going to compensate us for our forest? Or are you going to compensate us for our great river—for her fish, her water, for vegetables that grow along her banks, for the joy of living beside her? What is the price of this? How are you compensating us for fields either—we didn't buy this land; our forefathers cleared it and settled here. What price this land? Our gods, the support of those who are our kin—what price do you have for these? Our Adivasi (tribal) life—what price do you put on it? (Bava Mahalia, 1994, Letter from a tribal village)."

Here, according to O'Neill (2017, p. 1058), the problem with standard economic approaches to monetary valuation is not simple that this involves a narrow and inadequate consideration of backward- and forward-looking injustices or that it is cheaper to compensate the poor. There is also the matter of "[t]he dominance of a particular monetary language of valuation, and of the market order that gives this language of valuation its social power," which is itself a source of injustice as values are plural and incommensurable, with limited substitutability. As posited by Vatn and Bromley (1995, p. 9), "one metric (price) is unable to capture all relevant information" about the different kinds of values assigned to the environment because of the moral aspect of environmental choices. Arguments for wilderness and wildlife, for example, necessarily incorporate both economic and ethical considerations (Gowdy, 1997; Özkaynak et al., 2004). Therefore, there is no point in assuming that a particular ecosystem function or biodiversity loss, for instance, can be compensated by a gain in another function or in income.

It is at this point that the role of a public apology—rarely paid attention to today's environmental governance—might be worth noting. According to O'Neill (2017), an apology for sustainability policies is important as it comprises recognition of having committed a wrongful action and willingness to change the process and one's behavior in the future. It is an indication of responsibility not only in economic terms but also in political terms. Then, at the intersection of green economy and environmental justice, the questions posed by Martinez-Alier (2001, p. 153) become crucial: "Who has the power to impose particular languages of valuation? Who rules over the ways and means of simplifying complexity, deciding that some points of view are out of order? Who has power to determine which is the bottom line in an environmental discussion?"

Risk perceptions and the precautionary principle

Disagreements over the visions and policies best suited to securing a just and green future become especially visible when the struggle is between knowledge and risk in the face of scientific and technical uncertainties, and the recognition and participation claims (Martinez-Alier et al., 2010). Bridge et al. (2018) elucidates on how modern economic growth and ecological modernization discourse have given rise to environmental distribution conflicts and resulted in uneven geographies of risk and vulnerability. Meanwhile, Huber (2019) and McGoey (2012) underline that industries have increasingly taken advantage of uncertain evidence to ignore/hide claims of causal linkages—the so-called strategic ignorance. This was previously the case, for instance, with the public health impacts associated with asbestos, the tobacco industry, and the financial crisis; now the same thing is happening with climate skeptics (McGoey 2012; see also EEA, 2013). Today, the dominant prodam discourse celebrates hydropower as an uncomplicated, sustainable, and renewable source of energy, indispensable to development objectives such as green growth, climate change mitigation, and poverty alleviation (Schneider, 2013). Yet the critical accounts of Huber et al. (2016) and Huber (2019) highlight that the high-risk politics of hydropower in Italy and Spain as well as in the Eastern Himalayan region of Northeast India is mainly driven by powerful economic interests with neglect of the knowledge and safety of local communities.

As also argued by those involved in environmental justice struggles, uneven geographies of risk and vulnerability are not simply unintended consequences of growth but instead intentional and indeed a product of analytical and discourse-based narratives often based on expert knowledge systems and technocratic management decisions. This is in line with Kapp (1970), who indicates that environmental problems are not accidental effects of production and consumption, but rather systemic effects by dominant institutions. O'Connor (1998, p. 257) further explains this point: "the standard Western view that 'industrialization,' 'technology,' and so on are the causes of environmental destruction in both the West and the East fails to distinguish between a society's productive forces and its production relations, that is, its technological base, labor process, and production system, on the one hand, and its property, legal, and political relations, on the other."

Technology optimists often think that science and modern technology is precisely what will protect the environment and take us to a green economy, and therefore, pessimism about the long-term ecological effects of economic growth and certain technologies is unfounded. Conversely, Swaney (1987, p. 1749) states that "[t]o argue that technology will come to the rescue is to miss the scale of the problem, the importance of ecological systems for human wellbeing, the power of modern technology for harm as well as for good, the extent of our ignorance, or the limitations of modern western science." Hence, as Bullard (1994) and Faber (1998) put it, the struggle for environmental justice is not just about distributing risks equally but also about preventing threats from being produced in the first place. It is therefore important to encourage elaboration of "prudently pessimistic" policies that are often centered on the "precautionary principle" (Ravetz, 2004). The legitimacy of different value commitments is also important in finding appropriate ways of dealing with problems when uncertainties and social controversy make it difficult to be in full agreement about the facts (Faucheux and O'Connor 1998).

Currently, too little attention is given to policies based on technologically pessimistic assumptions, and global environmental governance is taking a "private turn" and gradually moving from precautionary, command-and-control state regulatory solutions to market-based solutions (Gareau and DuPuis, 2009, p. 2305). The extent to which the proponents of a green economy magnify the various voices that arise when researching risk and technology-related environmental policy issues and offer a distinctive "view from below" will be decisive in outcomes. As noted by Bridge et al. (2018) and Temper et al. (2018), histories originating from the experience of affected communities and environmental justice organizations can significantly enrich and transform policy analyses and outcomes. A green economy informed by the guiding principles of environmental justice definitely requires processes and procedures that are inclusive and transparent and bring together the range of information and viewpoints necessary for decision-making.

Temporal and spatial scales

A shared element of sustainability debates is the significance of intergenerational justice. Accordingly, present generations have a duty to protect at least the basic rights to survival, health, and subsistence of their children and of those yet to be born, and potentially their access to an environmental system that is of equal quality. As Howarth (2017, p. 258) underlines, "this is mainly based on the principle that an action is wrong if it benefits adults while impoverishing their living children and grandchildren, a straightforward conclusion is that each generation has a duty to ensure that life opportunities are sustained from one generation to the next." Such rights-based concerns are already present in IPCC reports: "If future people's basic rights include the right to survival, health, and subsistence, these basic rights are likely to be violated when temperatures rise above a certain level. However, currently living people can slow the rise in temperature by limiting their emissions at a reasonable cost to themselves. Therefore, living people should reduce their emissions in order to fulfill their minimal duties of justice to future generations" (IPCC Report, 2014, p. 216). Not much has been done so far.

Accepting that we are ethically obligated to future generations, the problem then becomes one of specifying a fair policy that will protect their rights. In fact, this issue is directly related to judgments about which natural and man-made resources are significant and essential for future generations. Norton and Toman (1997, p. 559) describe this as "the characterization of a bequest package" to leave for future generations. One approach, known as Hartwick's Rule (1977), interprets fairness as non-declining utility/consumption and considers an economy sustainable as long as investments in manufactured capital exceed the monetary value of natural resource depletion at all points in time (Howarth, 2017). In this sense, Solow (1986, 1993) indicates that since resources are fungible, intergenerational equity can be achieved through a fair investment policy. Thus, the current generation's obligation is mainly to leave a generalized productive capacity with at least the same standards of consumptive possibilities as today, for the future. As Howarth (2017, p. 259) underlines, the recent literature on the economics of climate change sheds light on the practical limitations of Hartwick's Rule in environmental governance in addition to the technical difficulties and/or theoretical challenges involved in using it.

In considering intergenerational equity, others such as Daly and Cobb (1989) and Costanza et al. (1996) support an opposing view, whereby they explicitly reject the idea of compensation for long-term environmental degradation by other consumption possibilities, and argue that to provide adequate options to future generations, natural capital and man-made capital stocks should be viewed separately. In other words, to ensure fairness across generations, "a structured bequest package" as Norton and Toman (1997, p. 559) put it is required, including stocks of natural resources that favor protection of large-scale processes and environmental quality, since there can be no real compensation for them in case of their loss. Note that this distinctive aspect of environmental justice also relates to issues of the "pace" and timeline of transition. In addition to the functional significance of natural capital for the ecosystem, this position also recognizes that natural capital can have symbolic or cultural significance for local communities that identify themselves with their habitats and landscapes (Noël and O'Connor 1998). Thus, the conservation of natural capital may, in this view, be motivated by the ethical and cultural conviction that there is no real substitute for these kinds of particular habitats. According to Howarth (2017), this approach still leaves room for decision-makers to rely on cost-benefit analysis and other valuation schemes, as long as attention is paid to the needs of future generations—mirroring the Kantian distinction between duty and prudence in the theory of rational action.

Efforts to implement environmental justice are further complicated by the international, national, and local scope of the concept. Pollutants not only "travel" from one place to another through multiple pathways and environmental processes, but they also traverse from one scale to another. At what scales do injustices occur, and how far do they reach? Where are decisions taken? Complexities arise especially at the level of scalar dynamics, since environmental governance decisions that appear more just or sustainable at one scale may be less so at a different scale (Heynen 2003). In this context, Siciliano et al. (2018), for instance, highlight how assessments of the priorities and needs, and hence of environmental justice implications, frequently diverge across different scales for affected groups. Using a multilevel energy justice framework, the authors show how energy decisions on infrastructure development can be taken based on energy justice principles and social impact evaluations. In a similar vein, Adaman and Devine (2017) point to the necessity of envisioning the local and the global as interdependent—the so-called glocal—processes in democratic governance structures, as global processes always involve some degree of localization, and local processes are part of a larger globalized web of networks.

Insights from the environmental and climate justice movements

As mentioned earlier, the concept of environmental justice provides communities and environmental activists with an important vocabulary in their resistance (Sikor and Newell, 2014; Schlosberg, 2007; Mohai et al. 2009). This is not a one-sided exchange, however. Insights from myriad local movements and international networks that have grown out of resource extraction and waste disposal conflicts at "commodity frontiers" are also important for framing the concept, since the concerns and claims articulated in these movements, while plural, are clear and consistent (Martinez-Alier, 2009; Martinez-Alier et al., 2016; Temper et al., 2015). Overall, the characterization of injustice used by movements addressed, as specified by Schlosberg and Collins (2014, p. 361), "distributive inequity, lack of recognition, disenfranchisement and exclusion, and more broadly, an undermining of the basic needs, capabilities, and functioning of individuals and communities." In this context, Low (1999, p. 6) referred to Harvey's account of the notes that "it is easier to find common ground in the shared experience of injustice than it is to move towards a shared understanding of justice." According to Agyeman (2014), the Principles of Environmental Justice developed by grassroots movements in 1991 offer a critical theoretical underpinning for merging class, race, gender, the environment, and social justice concerns. In addition, evidence presented by the Environmental Justice Atlas (Temper et al., 2015), an online database of environmental conflicts, also provides significant support for this observation, which is vital for a green economy eager for transformation in line with the guiding principles of environmental justice.

Moreover, as noted by Wright (2011), the environmental justice movement not only changed the very definition of the environment by criticizing the dominant Northern environmentalism that only cared about wilderness but also broadened the frame of environmental justice analysis for those in search of a green transition by drawing attention to the links between the ecosystem's health and the provision of justice, and the way environmental risks threaten everyday life. Accordingly, a green economy that is just necessarily involves "an understanding and supporting *both* Northern *environment-based* and Southern *equity-based* agendas, *equally*" (Agyeman, 2008, p. 181). Such a preanalytic vision then would disqualify any potentially narrow and inadequate interpretations of a green economy with major omissions and limited or false representations of the environment-society relationship.

Today, environmental justice movements do not only resist against the environmental manifestations of social injustice but also proactively engage in creating alternatives by constructing communities that are more just and sustainable. Such alternative imaginaries defy the image of society as composed of independent individuals, as is the case in mainstream economic thinking. They also necessarily go against the idea of the commodification of nature that turns ecosystem "services" such as fresh water, food, and carbon sequestration—that are free in principle, in terms of public or communal property—into commodities that can be acquired only by those who have purchasing power (Gómez-Baggethun and Ruiz-Pérez, 2011). Since justice and democracy requires interpersonal relations, a green economy vision inspired by environmental justice movements has to go much further than the vision discussed at Rio+20.

Insights from the climate justice movement are no different. Carbon offsets planned under clean development schemes are problematic, because most projects fail to deliver promised emission cuts. The "avoided emissions" are usually traded and exchanged for continued emissions in a developed country, and as such, commodification often results in no gains for the atmosphere and climate but a net transfer of rights and properties from the poor to the rich (Spash, 2007). According to Schlosberg and Collins (2014, p. 364), "[i]n an environmental justice approach, carbon markets are generally seen as giveaways to polluters at the expense of poor communities. A simple cap-and-trade system, where the original credits are given to polluters, is contrasted with preferred cap-and-dividend or fee-and-dividend policy, where permits would be auctioned to polluters and the revenue returned to poor and vulnerable communities. One activist explains the differing positions: 'Traditional climate activists espouse our economy works, except for the carbon thing. How do we simply make our economy less carbon-intensive?' But EJ folks see climate as a symptom of a whole system, so we need to rethink our economics." In this context, Ciplet and Roberts (2017, p. 149), referring to Okereke (2006), argue that "the dominance of *libertarian ideas of justice* have undermined distributive justice principles embedded in formative regime texts such as the United Nations Conference of Law of the Sea, the UNFFCC and the Basel Convention." Consequently, nowadays, responsibility for taking environmental measures seems to be viewed common to all, rather than on a "polluter pays" principle basis.

It should be noted that environmental movements of the South also insist on the existence of an ecological debt (including carbon debt). As Martinez-Alier (2002) clearly explains, their main point is not so much the calculation and payment of the debt but rather an acknowledgment of liability, and above all, a promise that the debt will grow no further. Yet, in Paris in 2015, the US chief climate negotiator Todd Stern denied compensation and liability. Payments for "loss and damage" were to be seen, according to him, not as compensation but as generous aid from rich to poor for coping with the effects of climate change when "adaptation" was not enough. Martinez-Alier (2015) noted, "[i]n fact money has been promised to poor countries if they give up their claims of a climate debt. Money is used for bribery."

A climate justice-based conception of a green economy, as is the case for environmental justice, requires looking even beyond distributive conceptions of justice and contemplating adaptation strategies that must take into consideration community voices, cultural impacts, and demands for capabilities that the community needs for sovereignty and functioning (Schlosberg and Collins, 2014; Turhan, 2016). Yet not much has been achieved so far. As Heffron and McCauley (2018, p. 76) point out, perhaps we need to return to the days when "[j]ustice was viewed to be applied if the protesting group were successful and they forced a legal and/or policy change to a project, i.e. there was a result. There needs to be a return to that clarity of thought with the just transition. More results are needed, i.e., two recent results for example from the UK in relation to a just transition is the phasing out of coal plants by 2025 and diesel cars by 2040."

Concluding remarks

The objective of this chapter was to comment on the role that environmental justice considerations can potentially play in a green economy and point to the possibilities and challenges facing the world in the environmental policy domain. The chapter made it clear that if green economy wants to go beyond rhetoric, technological optimism, and/or standard economic incentives, then it needs to embrace justice in all areas—social, environmental, and climate. While environmental justice might offer a comprehensive outlook for transitioning to a green economy, it more importantly serves to politicize conceptualizations of a green economy. More often than not, it is difficult to achieve unanimous agreement both on what constitutes environmental justice and what can be considered a green economy. In this context, the chapter highlighted that wisdom is in the details since differences in the way environmental justice is framed and implemented indeed impact policy outcomes in practice (Walker 2012).

As a result, it is crucial to carefully consider if, where, and how policies aimed at promoting a green economy consider all justice impacts and prospects. Critically discussing the different ways in which environmental justice is taken into account in a green economy setting renders some of the fundamental issues related to transitioning to a green economy debate transparent. The question then is who partakes in deciding what that transition looks like. Which definitions and whose notions of justice gain support in the public discourse, and whose voices are weak or left out? What kinds of social norms do we want to cherish, and ultimately, what kind of society do we want to live in? According to Spangenberg (2016, p. 127), "the different worldviews should become an issue of scholarly and public debate, as a choice between them will heavily influence the problems recognised and the policies derived to deal with them." Yet dialogue and informed participation by all affected parties can only be between equals, whereas the distribution of power in liberal and popular democratic systems is mostly skewed by short-term business and economic interests and needs. Regarding how to deal with negotiation failures due to asymmetries in bargaining power, Dryzek (1994, p. 194) noted that "[e]xactly how that might be accomplished without the heavy hand of the administrative state is a major unresolved issue, made especially problematic by the ability of capitalist and market systems to punish political decisions that impinge upon their logic of accumulation." In conclusion, as M'Gonigle (1999) rightly noted, the challenge is not so much a technical issue but rather a profoundly political and social one.

Unfortunately, as mentioned in the introduction, all this is nothing new. The same problems and solutions have been discussed and written about over and over again throughout the past decades in relation to sustainable development. The fact that many issues remain unresolved and are now being discussed again under the guise of a green economy is highly frustrating. Then, taking on Harvey's argument (1999), as underlined by Low (1999, p. 6), the best is "to regard ourselves as 'active agents caught within the web of life' and with the ability to change our political, social and (ultimately) ecological circumstances (at least those of our own making)." The time to act is now; academics and activists need to unite without delay, just as green economy and environmental justice should.

Acknowledgement

I gratefully acknowledge the financial support from the ACKnowl-EJ project [TKN150317115354] under the Transformations to Sustainability (T2S) programme coordinated by the International Social Science Council (ISSC) and funded by the Swedish International Development Cooperation Agency (Sida).

References

Aasen, M., Vatn, A., 2018. Public attitudes toward climate policies: the effect of institutional contexts and political values. Ecological Economics 146, 106−114.

Adaman, F., Devine, P., 2017. Democracy, participation and social planning. In: Spash, 2017.

Agyeman, J., Bullard, R., Evans, B., 2002. Exploring the nexus: bringing together sustainability, environmental justice and equity. Space and Polity 6 (1), 70−90.

Agyeman, J., 2008. Environmental justice and sustainability. In: Atkinson, et al. (Eds.), 171−188.

Agyeman, J., Evans, B., 2004. Just Sustainability': The Emerging Discourse of Environmental Justice in Britain? The Geographical Journal 170 (2), 155−164.

Agyeman, J., 2014. Global environmental justice or Le Droit Au Monde? Geoforum 54, 236−238.

Ballet, J., Jean-Marcel, K., Pelenc, J., 2007. Environment, justice and the capability approach. Ecological Economics 1, 1−8.

Baxi, U., 2016. Towards a climate justice theory. Journal of Human Rights and the Environment 7 (1), 7−31.

Bentham, J., 2009. An Introduction to the Principles of Morals and Legislation. Clarendon Press, Oxford.

Bentham, J.S., 1970. In: Burns, J.H., Hart, H.L.A. (Eds.), An Introduction to the Principles of Morals and Legislation. Athlone Press, London.

Boyce, J.K., 2004. Green and Brown? Globalization and the environment. Oxford Review of Economic Policy 20 (1), 105−128. http://oxrep.oupjournals.org/cgi/doi/10.1093/oxrep/20.1.105.

Brundtland, G.H., (chair), 1987. Our Common Future. Report of the World Commission on Environment and Development. Oxford University Press, Oxford.

Bridge, G., Barca, S., Özkaynak, B., Turhan, E., Wyeth, R., 2018. Towards a Political Ecology of EU Energy Policy. In: Foulds, C., Robison, R. (Eds.), Advancing Energy Policy. Palgrave MacMillan, pp. 163−175.

Bullard, R.D., 1994. Dumping in Dixie: Race, Class and Environmental Quality. Westview Press, Colorado.

Bullard, R., 1999. Leveling the playing field through environmental justice. Vermont Law Review 23, 454−478.

Cahn, C.M., 1949. The Sense of Injustice. New York University Press, New York.

Ciplet, D., Timmons Roberts, J., 2017. Climate change and the transition to neoliberal environmental governance. Global Environmental Change 46 (August), 148−156. https://doi.org/10.1016/j.gloenvcha.2017.09.003.

Costanza, R., Segura, O., Martinez-Alier, J. (Eds.), 1996. Getting Down to Earth: Practical Applications of Ecological Economics. Island Press, Washington, D.C.

Daly, H.E., Cobb, J.B., 1989. For the Common Good: Redirecting the Economy Toward Community, the Environment, and a Sustainable Future. Beacon Press, Boston.

Davidson, P.R., 2006. Just sustainabilities: development in an unequal world. Contemporary Sociology: Journalism Review 35 (2), 157–158.

Davies, A.R., Mullin, S.J., 2011. Greening the economy: interrogating sustainability innovations beyond the mainstream. Journal of Economic Geography 11 (5), 793–816. https://academic.oup.com/joeg/article-lookup/doi/10.1093/jeg/lbq050.

Dobson, A., 1998. Justice and the Environment: Conceptions of Sustainability and Dimensions of Social Justice. Oxford University Press, Oxford.

Dryzek, J.S., 1994. Ecology and discursive democracy: beyond liberal capitalism and the administrative state. In: O'Connor, M. (Ed.), pp. 177–197.

Dworkin, R., 2002. Sovereign Virtue the Theory and Practice of Equality. Harvard University Press.

EEA, 2013. Late Lessons from Early Warnings: Science, Precaution, Innovation. Copenhagen: European Environment Agency.

Faber, D. (Ed.), 1998. The Struggle for Ecological Democracy: Environmental Justice Movements in the United States. The Guilford Press, New York.

Faucheux, S., O'Connor, M. (Eds.), 1998. Valuation for Sustainable Development: Methods and Policy Indicators. Edward Elgar, Cheltenham, UK.

Fraser, N., Honneth, A., 2003. Redistribution or Recognition?: A Political-Philosophical Exchange. Verso Books.

Fraser, N., 2009. Scales of Justice: Reimagining Political Space in a Globalizing World. Columbia University Press, New York.

Gareau, B.J., DuPuis, M., 2009. From public to private global environmental governance: lessons from the Montreal Protocol's stalled methyl bromide phase-out. Environment & Planning A 41, 2305-232.

Gerber, J., et al., 2014. Socio-Environmental Valuation and Liabilities.

Gómez-Baggethun, E., Ruiz-Pérez, M., 2011. Economic valuation and the commodification of ecosystem services. Progress in Physical Geography 35 (5), 613–628.

Gowdy, J., 1997. The Value of Biodiversity: Markets, Society, and Ecosystems. Land Economics 73 (1), 25–41.

Hare, R.M., 1984. Rights, utility and universalization: a reply to Mackie, J.L. In: Frey, R. (Ed.), Utility and Rights. University of Minnesota Press, Minneapolis, pp. 106–121.

Hartwick, J.M., 1977. Intergenerational Equity and the Investing of Rents from Exhaustible Resources. American Economic Review 66, 972–974.

Harvey, D., 1999. Considerations on the environment of justice. In: Low, N. (Ed.), Global ethics and environment, London; New York: Routledge.

Heffron, R.J., McCauley, D., 2018. What is the 'just transition'? Geoforum 88 (August 2017), 74–77.

Heynen, N., 2003. The Scalar Production of Injustice within the Urban Forest. Antipode 35, 980–998.

Howarth, R., 2017. Future generations in Spash, Clive (2017). In: Routledge Handbook of Ecological Economics: Nature and Society. Routledge.

Hornborg, A., Martinez-Alier, J., 2016. Ecologically unequal exchange and ecological debt. Journal of Political Ecology 23 (1), 328–333.

Huber, A., et al., 2016. Beyond 'socially constructed' disasters: Re-politicizing the debate on large dams through a political ecology of risk. Capitalism Nature Socialism 5752, 1–21.

Huber, A., 2019. Hydropower in the Himalayan Hazardscape: strategic ignorance and the production of unequal risk. Water 11, 414.
IPCC, 2014. Climate Change 2014: Mitigation of Climate Change. In: Edenhofer, O., Pichs-Madruga, R., Sokona, Y., Farahani, E., Kadner, S., Seyboth, K., Adler, A., Baum, I., Brunner, S., Eickemeier, P., Kriemann, B., Savolainen, J., Schlömer, S., von Stechow, C., Zwickel, T., Minx, J.C. (Eds.), Contribution of Working Group III to the Fifth Assessment Report of the Intergovernmental Panel on Climate Change. Cambridge University Press, Cambridge, United Kingdom and New York, NY, USA.
IPCC, 2018. Global warming of 1.5°C. An IPCC Special Report on the impacts of global warming of 1.5°C above pre-industrial levels and related global greenhouse gas emission pathways. In: Masson-Delmotte, V., Zhai, P., Pörtner, H.O., Roberts, D., Skea, J., Shukla, P.R., Pirani, A., Moufouma-Okia, W., Péan, C., Pidcock, R., Connors, S., Matthews, J.B.R., Chen, Y., Zhou, X., Gomis, M.I., Lonnoy, E., Maycock, T., Tignor, M., Waterfield, T. (Eds.), the context of strengthening the global response to the threat of climate change, sustainable development, and efforts to eradicate poverty.
Kapp, K.W., 1970. Environmental disruption and social costs: A challenge to economics. Kyklos 23, 833–848.
Kant, I., 1998. Groundwork of the Metaphysics of Moral. Cambridge University Press, Cambridge.
Kenis, A., 2018. Post-politics contested: why multiple voices on climate change do not equal politicisation. Environment and Planning C: Politics and Space 0 (0), 1–18.
Kuehn, R.R., 2000. A taxonomy of environmental justice. Environmental Law Reporter 30, 10681–703. http://ir.lib.uwo.ca/cgi/viewcontent.cgi?article=1137&context=aprci.
Lele, R., 2013. Environmentalisms, justices, and the limits of ecosystems services frameworks. In: Sikor, T. (Ed.), The Justices and Injustices of Ecosystems Services.
Low, N., 1999. Global Ethics and Environment. Routledge, London and New York.
Lucier, Gareau, 2015. From waste to resources?: interrogating 'race to the bottom' in the global environmental governance of the hazardous waste trade. Journal of World-Systems Research 21 (2), 496–520.
Martin, A., et al., 2014. Whose environmental justice? Exploring local and global perspectives in a payments for ecosystem services scheme in Rwanda. Geoforum 54, 167–177.
Martin, A., et al., 2016. Justice and conservation: the need to incorporate recognition. Biological Conservation 197 (2016), 254–261.
Martin, A., McGuire, S., Sullivan, S., 2013. Global environmental justice and biodiversity conservation. Geographical Journal 179 (2), 122–131.
Martinez-Alier, J., 2001. Mining conflicts, environmental justice, and valuation. Journal of Hazardous Materials 86 (1-3), 153–170. http://www.sciencedirect.com/science/article/B6TGF-43V9RTJ-C/2/5c25c039d392da39b9812101713f4dd8.
Martinez-Alier, J., 2002. The environmentalism of the poor: a study of ecological conflicts and valuation. Edward Elgar, Cheltenham.
Martinez-Alier, J., Kallis, G., Veuthey, S., Walter, M., Temper, L., 2010. Social Metabolism, Ecological Distribution Conflicts, and Valuation Languages. Ecological Economics 70 (2), 153–158.
Martinez-Alier, J., 2009. Social metabolism, ecological distribution conflicts, and languages of valuation. Capitalism Nature Socialism 20 (1), 58–87. https://doi.org/10.1080/10455750902727378.

Martinez-Alier, J., 2015. Todd Stern, why don't you acknowledge the ecological debt? Ejolt blog. http://www.ejolt.org/2015/12/todd-stern-dont-acknowledge-ecological-debt/.
Martinez-Alier, J., Temper, L., del Bene, D., Scheidel, A., 2016. Is there a global environmental justice movement? Journal of Peasant Studies 43 (3), 732–755.
M'Gonigle, M.R., 1999. Ecological economics and political ecology: towards a necessary Synthesis. Ecological Economics 28, 11–26.
McGoey, L., 2012. Strategic unknowns: Towards a sociology of ignorance. Economy and Society 41, 1–16.
Meadows, D.H., Randers, J., Behrens, W.W., 1972. The limits to growth. Chelsea 205, 205. http://www.donellameadows.org/wp-content/userfiles/Limits-to-Growth-digital-scan-version.pdf.
Mill, J.S., 1967. Collected Works, vol.v. University of Toronto Press, Toronto.
Mohai, P., Pellow, D., Roberts, J.T., 2009. Environmental Justice. Annual Review of Environment and Resources 34 (1), 405–430.
Naess, A., 1990. Sustainable development and deep ecology. In: Engel, J.R., Engel, J.G. (Eds.), Ethics of Environment and Development: Global Challenge, International Response. The University of Arizona Press, Tucson, pp. 89–96, 1990.
Noël, J.F., O'Connor, M., 1998. Strong Sustainability and Critical Natural Capital. In: Faucheux, S., O'Connor, M. (Eds.), pp. 75–98.
Nussbaum, M.C., 2007. Frontiers of Justice: Disability, Nationality, Species Membership. Harvard University Press, Cambridge, MA, and London.
Norton, B., Toman, M.A., 1997. Sustainability: ecological and economic perspectives. Land Economics 73 (4), 553–568.
O'Neill, J., 2017. The price of an apology: justice, compensation and rectification. Cambridge Journal of Economics 41, 1043–1059. https://academic.oup.com/cje/article-abstract/41/4/1043/3859793. July 27, 2018.
Okereke, 2006. Global environmental sustainability: intragenerational equity and conceptions of justice in multilateral environmental regimes. Geoforum 37, 725–738.
Okereke, C., Ehresman, T.G., 2015. International environmental justice and the quest for a green global economy: introduction to special issue. International Environmental Agreements: Politics, Law and Economics 15, 5–11. https://doi.org/10.1007/s10784-014-9264-3.
Özkaynak, B., Devine, P., Rigby, D., 2004. Operationalising strong sustainability: definitions, methodologies and outcomes. Environmental Values 13 (3), 279–303.
Pellow, D.N., 2001. Environmental justice and the political process: Movements, corporations, and the state. Sociological Quarterly 42 (1), 47–67.
Pulido, L., 2017. Conversations in environmental justice: an interview with David Pellow. Capitalism Nature Socialism 28 (2), 43–53. https://www.tandfonline.com/doi/full/10.1080/10455752.2016.1273963.
Ravetz, J., 2004. The post-normal science of precaution. Futures 36 (3), 347–357.
Rawls, J., 1971. A Theory of Justice. Oxford University Press, Oxford.
Redclift, M., 2018. Sustainable development in the age of contradictions. Development and Change 49 (3), 695–707. http://doi.wiley.com/10.1111/dech.12394.
Rockström, J., et al., 2009. Planetary boundaries: exploring the safe operating space for humanity. Ecology and Society 14 (2), 472–475. http://www.nature.com/nature/journal/v461/n7263/full/461472a.html.

Rodríguez, I., Sletto, B., Bilbao, B., Sánchez-Rose, I., Leal, A., 2013. Speaking of fire: reflexive governance in landscapes of social change and shifting local identities. Journal of Environmental Policy and Planning 1–20.

Schneider, H., 2013. World Bank Turns to Hydropower to Square Development with Climate Change. Available online: https://www.washingtonpost.com/business/economy/world-bank-turns-to-hydropower-to-square-development-with-climate-change/2013/05/08/b9d60332-b1bd-11e2-9a98-4be1688d7d84_story.html?utm_term=.1406f957166c. (Accessed 4 July 2019).

Schlosberg, D., 2007. Oxford University Press Defining Environmental Justice: Theories, Movements, and Nature. Oxford University Press, Oxford.

Schlosberg, D., 2013. Theorising environmental justice: the expanding sphere of a discourse. Environmental Politics 22 (1), 37–55. http://www.tandfonline.com/doi/abs/10.1080/09644016.2013.755387.

Schlosberg, D., Collins, L.B., 2014. From environmental to climate justice: climate change and the discourse of environmental justice. Wiley Interdisciplinary Reviews: Climatic Change 5 (3), 359–374.

Scoones, I., 2016. The Politics of Sustainability and Development. Ssrn.

Sen, A., 2009. The Idea of Justice. Harvard University Press Cambridge, Massachusetts.

Siciliano, G., Urban, F., Tan-Mullins, M., Mohan, G., 2018. Large dams, energy justice and the divergence between international,national and local developmental needs and priorities in the global South. Energy Research and Social Science 41, 199–209.

Sikor, T., 2013. The Justices and Injustices of Ecosystem Services. Earthscan, London.

Sikor, T., 2016. "Rethinking Environmentalism." (Hackmann 2013).

Sikor, T., Martin, A., Fisher, J., He, J., 2014. Toward an empirical analysis of justice in ecosystem governance. Conservation Letters 7 (6), 524–532.

Sikor, T., Newell, P., 2014. Globalizing environmental justice? Geoforum 54, 151–157.

Solow, R.M., 1986. On the Generational Allocation of Natural Resources. Scandinavian Journal of Economics 88, 141–149.

Solow, R.M., 1993. Sustainability: An Economist's Perspective. In: Dorfman, R., Dorfman, N. (Eds.), Selected Reading in Environmental Economics. Norton, New York, pp. 179–187.

Stevenson, H., 2015. Contemporary Discourses on the Environment–Economy Nexus. SPERI Paper No. 19, ISSN 2052-000X, available at: http://speri.dept.shef.ac.uk/wp-con- tent/uploads/2015/03/SPERI-Paper-19-Contemporary-Discourses-on-the-Environment- Econ omy-Nexus.pdf.

Spash, C.L., 2007. The economics of climate change impacts ?? La stern: novel and nuanced or rhetorically restricted? Ecological Economics 63 (4), 706–713. http://linkinghub.elsevier.com/retrieve/pii/S0921800907003497.

Swyngedouw, E., 2011. Interrogating post-democratization: reclaiming egalitarian political spaces. Political Geography 30 (7), 370–380. https://doi.org/10.1016/j.polgeo.2011.08.001.

Temper, L., et al., 2018. A perspective on radical transformations to sustainability: resistances, movements and alternatives. Sustainability Science 0 (0), 0. http://link.springer.com/10.1007/s11625-018-0543-8.

Temper, L., Daniela del Bene, Martinez-alier, J., 2015. Mapping the frontiers and front lines of global environmental justice : the EJAtlas. Journal of Political Ecology 22, 256–278.

Turhan, E., 2016. Value-based adaptation to climate change and divergent developmentalisms in Turkish agriculture. Ecological Economics 121, 140–148.

Vatn, A., Bromley, W.D., 1995. Choices without Prices without Apologies. In: Bromley, D. (Ed.), Handbook of Environmental Economics. Blackwell, UK, pp. 3–26.

Walker, G., Bulkeley, H., 2006. Geographies of environmental justice. Geoforum 37 (5), 655–659.

Walker, G., 2012. Environmental justice: concepts, evidence and politics. Routledge, London and New York.

Whiteman, G., 2009. All my relations: understanding perceptions of justice and conflict between companies and indigenous peoples. Organisation Studies 30, 101–120.

Wright, N.G., 2011. Christianity and environmental justice. CrossCurrents 61, 161–190.

Zografos, C., Rodríguez-Labajos, B., Aydin, C.A., Cardoso, A., Matiku, P., Munguti, S., O'Connor, M., Ojo, G.U., Özkaynak, B., Slavov, T., Stoyanova, D., Živčič, L., 2014. Economic tools for evaluating liabilities in environmental justice struggles. The EJOLT experience. EJOLT Report No 16, 75p.

CHAPTER

7

Balancing climate injustice: a proposal for global carbon tax

Rohit Azad, Shouvik Chakraborty

Introduction

Climate scientists have pressed the panic button. They are warning that time is running out. A recent study published by the Proceedings of the National Academy of Sciences shows that if immediate actions are not taken, then the earth will permanently change into a "Hothouse Earth," beyond which any efforts made to reverse these developments will prove futile because that threshold limit will have been crossed (Steffen et al., 2018).

What is the time frame? The most recent Intergovernmental Panel for Climate Change (IPCC) report suggests that we, as humankind, might have just over a decade left to limit global warming. However, this is only possible if there are rapid, far-reaching, and unprecedented changes in all aspects of our lifestyles, especially those of the affluent in our societies. And that is because, as the IPCC report shows, there is more than a 95% probability that human activities are responsible for warming the planet and climate change. To stabilize the temperature to no or limited overshoot of 1.5°C, total global emissions will need to fall by 45% by 2030 and reach net zero by 2050. If these targets are not met, tropical regions of the world, which happen to be particularly concentrated and densely populated in the global South, are likely to be most negatively affected because of their low altitudes and preexisting high temperatures (Mendelsohn et al., 2006; Martin, 2015). Martin (2015) argues, based on a report by a United Kingdom−based risk analysis firm Maplecroft, that of the top 32 countries at "extreme risk" from climate change, the top 10 are all tropical countries.

There is a clear paradox here if we divide the world into two halves: global South and global North. The global South, which has historically contributed less to the problem (and even currently their per-capita carbon emissions are much lower in comparison with the countries in the global North), happens to be at the receiving end of the lifestyle choices made by the global North. A singular focus of this chapter is to address this climate injustice at the global level.

Unfortunately a genuine consensus on mitigation of this problem is missing. In the global North, the argument is that a large part of the problem lies with the global South, especially with rapidly growing countries such as China and India, the effect

Handbook of Green Economics. https://doi.org/10.1016/B978-0-12-816635-2.00007-9

of which is apparent in their total emissions being among the highest in the world today and in certain instances more than some countries in the global North. Given that is the case, so goes the argument, the burden of adjustment needs to be shared equally between the North and at least these high carbon–emitting countries in the South.

In the South, on the other hand, these are usually presumed to be the "problems of the North." A large number of articles, ideas, and policy recommendations on this issue of environmental degradation that are coming from the respective parts of the world capture the sense of this difference. This dichotomy between the two positions on a pressing problem such as the environment and the relayed issue of climate change is, however, not only counterproductive but also imprudent toward our future in many ways.

There is absolutely no doubt that historically the global North has contributed the most to the carbon *stock* that has built up in the environment, which has led to the process of global warming, a process that can be traced to as far back as the Industrial Revolution. So, these advanced countries in the North cannot absolve themselves by simply hiding behind current high emissions by certain countries in the global South. The North, therefore, will have to bear a greater burden of the adjustment for the stock of carbon emitted historically over the past two centuries. In other words, sharing of this burden cannot be based on the current flows alone but should take into account the stock of carbon contributed in the past.

At the same time, the *flow* of emissions is also important since these flows after all add to the carbon stock existing in the environment. Similar to the North's contribution to the carbon stock, it is also beyond doubt that the emerging market economies are contributing a decent share to the flows, albeit in absolute terms and not per capita terms. Moreover, for the global South, when it comes to the environment, the argument simply cannot be that "since you have polluted earlier, we have a right to do so today" or that by asking for a blueprint of gradually declining emissions targets, the ladder is being taken away from the emerging market economies. An interesting component of the proposal we make here is that growth is not juxtaposed with sustainability. In fact, as we discuss later, they can go together, which is a win-win situation for the global South as well as for the world as a whole.

In a more obvious way, it is only with a collective effort from the North and the South can this serious problem of climate change be stalled. And given that the global South has more to lose in the immediate sense because of their tropical location, it is imperative on them to look at the problem more objectively.

There is no denying that there is a huge gap between the standard of living of the people in the North and those of the South for historical as well as contemporary reasons. As a by-product of that very process is the huge gap between per capita emissions in these two worlds. This just shows such climate injustice is inherent in the current capitalist growth process. Since the growth process has historically generated two poles, at the one end of which is opulence with penury at the other, a fair response to this problem would be to bridge this huge gap. This gap needs to be bridged for sure because it will be unfair to expect one section of the global population to live a life with little dignity, whereas the other section enjoys all the profligacy. But bridging this gap should not be at the cost of the environment either. Is there a third route possible, which keeps growth, sustainability, and equity in balance? Our proposal answers this question in the affirmative.

While the gap in standard of living of an average citizen between the two worlds is huge, hiding behind low per capita emissions even as absolute emissions are high is perhaps akin to acting like an ostrich—"hiding behind the poor." What matters for the world is not the incremental contribution of an individual to total emissions but what we, as a humankind, are contributing to the environment. Once the temperatures rise globally, it will not choose its victims based on their individual contribution to the problem. In fact owing to the poor socioeconomic conditions, high population density along coastlines, job dependence on agriculture and other allied activities in the South, and the casualties of environmental degradation (arising out of rising water levels, cyclones, and other devastating natural events) are going to be significantly more than the North.

So, both the worlds need to contribute to avert this danger in their self-interest. At the same time, the burden of adjustment *cannot* be equal when the underlying relationship between the two worlds has been unequal for all these years. But what is the correct balance in terms of sharing this burden, something that can be politically and juridically just?

A just approach would involve a global sharing of the burden between countries according to their respective shares in global emissions. Here is the gist of the proposal that is discussed below. Based on this idea of climate justice, we calculate a global carbon inequality quotient (GCIQ), which measures the ratio of a country's contribution to global emissions and their share in the global population. A comprehensive green energy transition for the world requires, according to most estimates, 1.5% of each country's gross domestic product (GDP) over a period of 20 years (Pollin et al., 2015; Pollin and Chakraborty, 2015). Our proposal, JET (a Just Energy Transition), requires countries contributing less than the global average to mobilize a part of their resources through carbon taxes commensurate with their distance from the average and the deficit billed by countries with per capita emissions greater than the global average, who will consequentially have to mobilize more than their own transition requirements. A part of the tax would be distributed in the form of universal carbon dividends to keep the poor insulated from the regressive impact on their income from this policy. Most of the climate change negotiations continue to deal with how much of the burden is to be shared between the North and the South. In the absence of a common agreement on this, environment has become the casualty. Our proposal in contrast will not only be a historically just way to balance the environmental and ecological damages done in the past but also help the resource-poor developing countries to make the energy transition without having to unduly worry about the finances.

Carbon emissions and current strategies

To understand the different strategies through which emissions can be reduced, a simple diagram can help. Let us assume that economic activities require emitting carbon into the environment even though the extent of these emissions would depend on the fossil fuel dependence as well as efficiency of its usage for a particular activity. Assuming, like in much of economics, that there is a marginal product of carbon (input) declines as more is emitted, this gives rise to a downward sloping curve where marginal benefit (MB) is measured on the y-axis and carbon emissions on the x-axis. A downward sloping MB curve measures the benefit (revenue) that a firm gets by emitting one more unit of carbon into the environment. MB declining is not an unreasonable assumption since the underlying technology is taken to be given and the other inputs are considered to be fixed in supply in the short run. Once these assumptions are relaxed, the shape as well as position of the curve also changes. Every economic activity has such an MB curve, and a weighted average of this across the different activities and sectors in the economy will give us an economy-wide MB curve (Fig. 7.1).

In most instances, there is no price attached to emitting carbon, even though this is a negative externality for the people in general. In the absence of any carbon price,

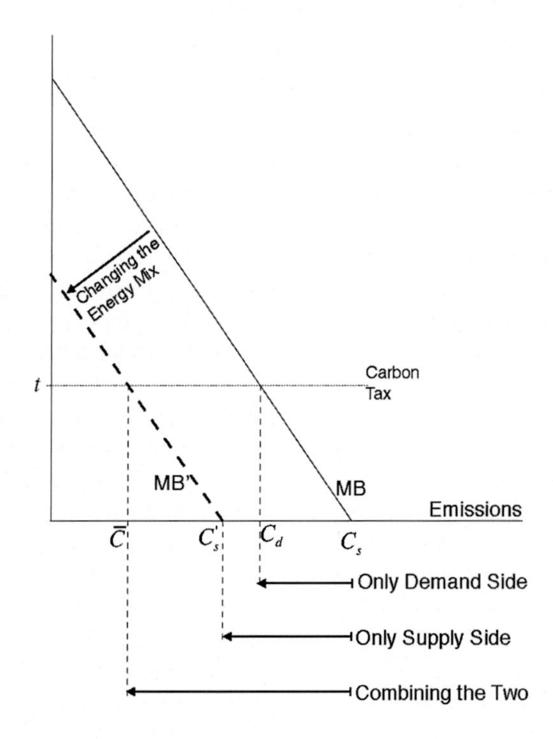

FIGURE 7.1

Impact of the Carbon Tax and Changing Energy Mix on Emissions explicit or implicit, a market economy driven purely on considerations of profits will emit C_s amount of carbon since that is the point at which the MB curve meets the zero marginal cost of carbon.

The purpose of environmental policies is to decrease these emissions according to preset targets to control rising temperatures in the economy. Let us say that the carbon needs to be restricted to C_d level. Given that the MB curve is not known with certainty, different variations of environmental policies would basically be constructing an *expected* MB curve. Now for a given MB, C_d can be achieved either by fixing quantity at C_d or price t. Fixing one means the other will be freely determined depending on the *actual* position of the MB curve.

The mitigating strategies can broadly be classified into three categories, which can be visualized at the local, national, or even global level:

1. Regulatory (indirect) mechanisms: Regulatory mechanisms such as setting performance standards, energy efficiency labels on products, and targeted subsidies are measures that may have an effect on emissions through a voluntary shift either by the consumer or the producer toward greener forms of products/processes.
2. Carbon targeting (direct): A given amount of emission can be targeted either by capping it at a level, which is the most effective way of controlling carbon, or by levying a tax on it, which through the price channel affects the level of emissions. Boyce (2018) argues that carbon pricing is a key instrument in climate policy since it creates incentives for cost-effective emission reductions in the short run and cost-reducing innovation in the long run. Carbon capping can be achieved in two ways: cap-and-trade and cap-and-giveaway. The former requires the regulatory (local, national, or even global) authority to fix the emissions at the certain level and issue equivalent permits, which are then auctioned in the market, which sets the price of carbon. The latter involves fixing of the target, but the permits are given free to firms according to some criteria of their past emissions record or their adoption of greener forms of technology. When a carbon tax or cap-and-trade (though this doesn't hold for cap-and-giveaway), the price of fossil fuel rises and with it the prices of all other commodities. However, the rise is proportionate to the carbon embodied in these commodities. Any such rise in carbon price, however, is regressive in its redistributive effects since the poor consume a higher proportion of their income and hence the incidence of a commodity tax is more likely to affect them per unit of income as opposed to the rich. This inbuilt injustice of a carbon tax, where the burden of adjustment falls disproportionately on those contributing less to the problem, can be addressed either through cap/tax-and-distribute, where the carbon revenue is distributed evenly across all or cap/tax-and-spend, where the revenue is used to subsidize the cost of living of poorer sections. The distributive impacts of such a policy in the US context has been discussed in great details by Boyce (2018), Boyce and Riddle (2007), and Fremstad and Paul (2017) and in the Indian context by Azad and Chakraborty (2018).

3. Changing the energy mix (direct): The third mechanism, which shifts the MB curve inward is an effective mechanism to counter climate change. This has been discussed in great details by Pollin (2015). How does the MB curve move in? If more activities in the economy are moving toward greener forms of technology, which have no emissions (or very little emissions, perhaps at input level stages), the weighted average of the curves will be closer to the origin with the extreme case of complete disappearance of the curve once fossil fuel usage has been completely substituted for.

Each of these three strategies has pros and cons. Regulatory mechanisms may have limited success since they depend, as a result of their indirect nature, to a large extent on the voluntary adoption by the producers and consumers alike. Moreover if the targeted subsidies are in the wrong and inefficient alternatives, then that harms the energy transition irreversibly. Carbon pricing, on the other hand, whether as a cap or tax, limits emissions more directly than the first mechanism; hence, it is more effective. But given that, on its own, carbon targeting can at best fix a specific point on the MB curve but not shift the MB curve inward, it is more of a short- to medium-run solution, albeit a powerful one within that time frame. The third mechanism, although very effective, has a longer gestation lag and requires a fundamental change in the production structure of the economy, which under normal circumstances has a lot of inertia because of competitive and technological reasons.

An ideal policy mix would, accordingly, be one that combines carbon pricing with a target-based supply-side response. So, while in the short run, the economy moves up the MB curve, over time, the MB curve itself shifts inward. So, a combination of the movement along as well as of the MB curve can help bring the emissions down. We have shown such a possibility in Fig. 7.1. If the policymakers rely only on carbon pricing policy, emissions come down to C_d, and if they exclusively focus on supply-side policies, the emissions come down to C'_s (since the marginal cost of emitting carbon in the absence of a tax is zero). A combination of the two policies will bring the emissions down to $\overline{C}$.

Since the concern of this chapter is at the global level, each of these strategies can be visualized in an appropriate context. At the global level, the most accepted model of mitigating strategy has been the carbon trading process. What this process does is, in terms of the MB curve (in this case seen at the global level), it fixes a target per capita level of global emissions, let us say C_d, and those to the left of this level can sell carbon credits to those on the right of this level. And the price of the permit is determined by the relative demand and supply for these permits by the respective countries.

While this may lead to a fall in emissions, and that too if globally accepted and properly implemented, it has its own limitations. First, the price of carbon determined in this manner is going to depend on the bargaining power of the two sets of countries with the wealthy and the powerful more often than not setting the price of these permits. So, it is quite possible, like in the case of many negotiations in the WTO, that the global South will get the short end of the stick. Second, it effectively puts the burden of climate adjustment on those contributing less to the problem. So, even though they might get compensated in terms of revenue through these permits, unless the permits are justly priced, it may not be enough to make the energy transition in their own countries, thereby keeping their production and energy structures intact. This could be both because of a lower revenue than what is required to make the energy transition and/or lack of access to technology. In such a scenario, the global North will continue to prosper at the cost of global South since the revenues for the latter will come only in so far as you continue to be to the left of the target point, i.e., at a lower level of economic development.

As happens in the case of any international negotiation, whether of borders or commodities or technology, countries that are more powerful are the ones who determine the terms of bargain and till these underlying power relations are changed, any so-called market determined prices do not represent just-and-efficient prices. So, this will perpetuate the inequities that have existed between the two worlds even though emissions might come down. We ask the question whether there can be a policy that does the opposite, i.e., brings the emissions down but in the process lowers the inequities instead of reinforcing them?

Justice, based on hard objective facts, is better than efficient-sounding-but-unjust market-based outcomes, which is what the current carbon trading process entails. No wonder that the poorer countries feel cheated and find this process to be perpetuating their underdevelopment. What could this alternative proposal be?

Our proposal

Since our proposal is premised on some sense of global justice in terms of climatic fallouts and the respective contributions of the countries, it would be appropriate to present a picture of what this climate injustice currently looks like. We have divided countries into five income categories such as high, upper middle, middle, lower middle, and low and plot their per capita emissions against their income. Fig. 7.2 presents a very stark picture of the inequality in carbon footprints between the richer and the poorer nations. That everyone is aware of the inequality is one thing but to get such a clear "funnel" function shows that those at the bottom who contribute the least are the one who are going to suffer the most as a consequence of actions of those at the top. So, that is double jeopardy for the global poor without even being responsible for it! Any climate mitigation strategy has to be in a direction that rights this unacceptable wrong committed by the global rich.

We also discussed the concept of the GCIQ, which is a measure of the ratio of the proportion of CO_2 emitted by a country in the total global emissions to their share in the global population. It simply measures the share of the carbon footprint of classes as a proportion of their share in the population, which shows that of 203 countries studied, 131 countries emit lesser carbon than their share in the population ($CIQ < 1$) and the remaining 72 countries emit more than their share in the population ($CIQ > 1$). Among the latter, the major polluters are the oil-producing

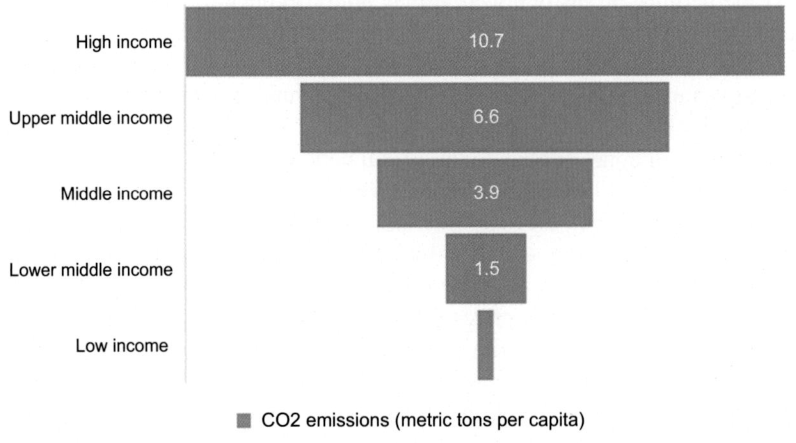

FIGURE 7.2

The relationship between level of income and level of emissions.

Source: Authors' calculations based on data from World Development Indicators, World Bank.

countries, such as Qatar (9.71), the UAE (4.98), Saudi Arabia (4.17), and Kuwait (5.39), and the rich nations, such as the United States (3.52), Canada (3.23), and Australia (3.28) (Fig. 7.3).

How can this injustice be corrected while making the planet a better place to live in for the future generations? The first priority is a fundamental change in the energy infrastructure, which requires massive investments for the green energy program across the world. What we propose here in some sense is a new global green deal. But how is it to be financed? It is here that the role of the climate justice kicks in. So, we propose that those on the top of the funnel, apart from funding their own energy transition, partially fund the transition for the countries at the bottom and this sharing of the burden of transition be done in a way that inverts the injustice funnel. Let us look at this proposal in some detail.

It has been shown, as mentioned earlier, that to make a successful energy transition from fossil fuels to greener renewable sources, countries have to spend around 1.5% of their GDP. Of course, this level would vary with countries depending on their distance from what can be called the green frontier, i.e., the most green energy–dependent country of the time.[1] So, those who are further away from the frontier will have to spend more than 1.5% and those closer may have to spend less. As a first approximation, we could say that on an average, if 1.5% global

[1] As Pollin (2015) shows that some countries such as Brazil and Germany can spend at lower levels of 1.0%.

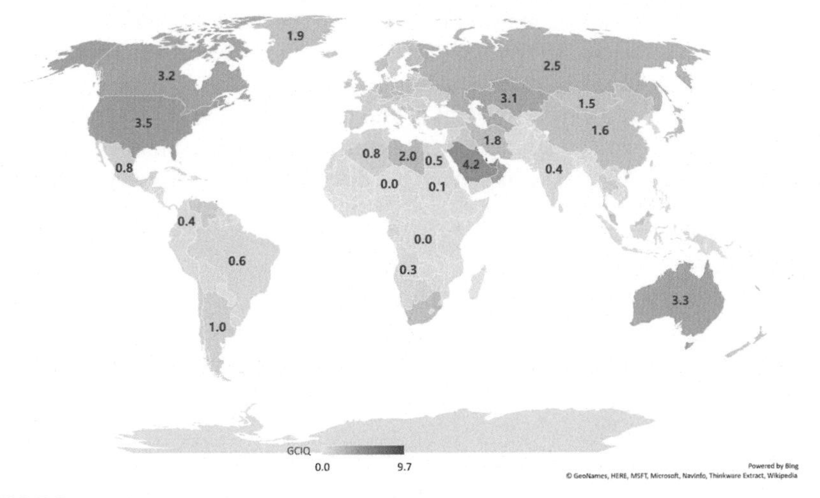

FIGURE 7.3

The global carbon inequality quotient.

Source: Authors' calculations based on World Development Indicators, World Bank.

GDP is spent on the energy transition over the next two decades, we will be able to avert the danger that the climate scientists have been warning against.

This amount gives us the total revenue required at the global level to make the transition. In terms of Fig. 7.1, this is the level of the expenditure required to gradually shift the MB curve inside, at the global level, over the next two decades. The immediate question that arises is where would this money come from? One option is that the respective governments fund this from their own treasury. There are three problems with this, particularly from the perspective of the resource-poor countries. First, they might not have the fiscal headroom to make this expenditure, especially if they are trade-dependent economies. Second, such a country-specific policy places the mitigation responsibility *equally* on each country irrespective of their contribution to the problem. Third, such a policy because of fiscal conservatism in many countries may not even see the light of the day let alone change the infrastructure. It is here that carbon tax enters the picture. We propose that the global energy transition is financed through a system of global carbon tax. The policy of course keeps the option open for countries to finance their respective expenditures purely by running a fiscal deficit and not take the route of carbon tax necessarily. But since carbon tax has an additional impact of limiting emissions, a policy that combines the supply and demand side would be a better mitigating strategy. For the exercise below, we assume that all countries levy carbon taxes albeit the effective tax rates, i.e., after adjusting for the carbon tax transfers from the top of the funnel to the bottom, vary across countries according to their position on the climate injustice funnel in Fig. 7.2.

Who subsidizes whom and by how much? The benchmark is drawn at the current global average per capita level of emissions. Those countries emit more than that pay for their own transition plus fund a part of the transition of those who are below this average. So, those at the receiving end of climate injustice are duly compensated for even as the entire world transitions to a greener earth as a result of this process of carbon tax sharing. The exact nature of the distribution of taxes is given below.

Methodology

Estimating the universal global carbon tax

We assume that the total level of CO_2 emissions in the world economy is X.[2] Let us also denote the world population as N. Then, we can state that the global per capita emissions is $\overline{X}$, which is simply the ratio of total CO_2 emissions to the population (i.e., $\overline{X} = X/N$).

Let there be n number of countries in the global economy, where x_i represents the per capita emissions for the ith country and N_i denotes the population of the ith country ($\forall i = 1(1)n$). The total amount of funds required per year to improve the energy efficiency of the global economy and generate clean, renewable energy, be denoted as R.[3] To achieve this goal of energy transformation, we propose, as discussed above, a global carbon tax collected by each country or nation state. However, the collection of this tax involves some administrative cost, which is covered within the collected taxes. In the literature, it is usually assumed that the administrative costs are approximately 1% of the collected taxes. We assume that $\widehat{R} = 1.01 * R$ denotes the amount of total tax revenue generated including the administrative costs. Suppose the world decides to levy a uniform tax rate globally to mobilize the resources required to invest in this clean energy transformation, and then the uniform tax rate levied per unit of metric ton of CO_2 will be $t\left(= \left(\widehat{R}/X\right)\right)$. Then, one can argue that the total individual tax collected by the ith country at this uniform rate will be $T_i = t^*x_i^*N_i$ ($\forall i = 1(1)n$). Therefore, it means that $\widehat{R} = \sum_{i=1}^{n} T_i$.

Estimating the tax of the "payers"

However, given the historical disparity in the amount of energy usage and the level of developments as argued earlier, it would be an unfair process if all countries are equally taxed. We define a simple rule of distribution. The countries where per capita emissions are more than the global average, the "payers," transfer a part of their

[2] We consider the level of CO_2 emissions only since it is emitted primarily through the usage of fossil fuels. There are, however, other greenhouse gases such as methane (CH_4), nitrous oxide (N_2O), and fluorinated gases that also contribute to global warming.

[3] The annual budget for clean, renewable energy transformation is calculated as 1.5% of the average of the GDP in the base year (0) and the projected GDP in the final year (t), i.e., $R = 0.015^*(GDP_0 + GDP_t)/2$.

carbon tax revenue to a global, international body, which distributes these collected taxes among those countries, the "beneficiaries," which are polluting less than the global average in per capita terms. The amount that the "payers" pay depends on the difference between their per capita emission levels and the global per capita emission levels times their population. Now, suppose we denote the amount of transfers made by the ith country to be S_i. Then, the transfer, S_i, made by the ith country, will be $S_i = (x_i - \overline{x})^* t^* N_i$, where N_i denotes the population of the ith country $(\forall i = 1(1)k)$ and k denotes the number of nations whose emissions are above the global average. Hence, the total transfers made by these k countries can be written as $S = \sum_{i=1}^{k} S_i$.

These k countries, which we are calling "payers," also need resources for their own energy transformation to achieve the global emission goals. Suppose the amount of resources needed for the energy transformation of these individual countries be denoted as R_i, where $i = 1(1)k$. The amount of total taxes retained by these economies after the transfer of payments are $\overline{T}_i = \overline{x}^* t^* N_i$ $(\forall i = 1(1)k)$. Now, if $\overline{T}_i > R_i$, then these individual economies can lower their tax burdens. However, if $\overline{T}_i < R_i$, then these countries need to mobilize additional resources to achieve the domestic energy transformation goals. It can achieve these goals either through deficit-financing or mobilizing more taxes. To estimate the actual amount of taxes, we assume that these economies take the latter route.

Estimating the tax of the "beneficiaries"

How does the total sum S received from the payer countries get redistributed among the $(n-k)$ "beneficiaries," which will receive a portion from those payments? We give emphasis on two important determinants: (1) the less these countries pollute, the more they are eligible for resources and (2) the larger the population of the countries, implying as a proxy the larger size of the economy, and hence larger share from the payers' contributions. We define the weights in each of these categories as w_{1i} and w_{2i} $(\forall i = 1(1)\ (n-k))$. Therefore, based on these weights, the amount (A_i) received by these individual economies $A_i = \left(w_{1i}{}^* w_{2i} \Big/ \sum_{i=1}^{(k-n)} \left(w_{1i}{}^* w_{2i} \right) \right)^* S$.

Now, these economies have mobilized their carbon taxes $\widetilde{T}_i = x_i{}^* t^* N_i$ $(\forall i = 1(1)\ (n-k))$ at the universal tax rate applied for all the economies. Hence, the total resources available to spend for transformation of the energy system and to improve their access will be $TR_i = A_i + \widetilde{T}_i$ $(\forall i = 1(1)\ (n-k))$.

If these total resources mobilized from their own domestic tax collection and the transfers received from the "payer" countries are higher than those needed for the transformation to clean, green energy, i.e., $\left(\widetilde{T}_i + A_i\right) \rangle R_i \left(\forall i = 1(1)(n - k)\right)$, then the resources can be used to create the energy infrastructure required for access to basic facilities such as electricity or equally distribute those resources in "cash" or "kind" among the individuals. The other alternative is those economies can choose to lower the tax rates. However, if the reverse condition holds true, i.e., $(\widetilde{T}_i + A_i) < R_i(\forall i = 1(1)(n - k)))$, then these economies, similar to the payer countries, need to mobilize the additional resources either through further increase of the tax rate or through deficit financing. We estimate the tax rates for these countries, assuming for both the cases, it is done through adjusting the tax rates of these individual economies.[4]

Data sources

The data source for this chapter is the World Development Indicators, World Bank, which has data on carbon emissions, population, and energy access till 2014. Due to the limitation of the projected database on the GDP of individual countries, we had to rely on the Organisation for Economic Co-operation and Development (OECD) series—the long-term baseline projections. For this reason, we could not cover all the countries of the globe, except those which are in the OECD data series. We used the projected GDP of 2035 for these individual countries from the database.

Results

The three broad categories of expenditures required for global green energy transformation are clean renewables, energy efficiency, and administrative costs with respective amounts of USD 1.1 trillion, USD 550 billion, and USD 16.5 billion. Since the global carbon emissions is 36.1 billion metric ton of CO_2, this amounts to a global carbon tax of USD 46.1 per metric ton. This is the universal global carbon tax that, we assume as a start off, needs to be levied across the world to finance the global energy transformation. A uniform carbon tax globally makes the possibility of arbitrage impossible.

While a uniform global carbon tax may look good on paper, it would simply be unacceptable since this means equating all countries at par when it comes to the mitigation problem. It would mean severe hardships especially for the poorer sections in the poorer regions of the world because of a substantial price rise, resulting from these carbon taxes.

The next step, therefore, is to calculate the way in which this carbon burden can be shared among various economies. The global average of carbon emissions is 4.97 metric ton per capita. All the countries with emissions above this level (68 in all) are "payers" to finance energy transition for "beneficiary" countries, 135 in number, which are emitting below this level.

The total amount of "carbon compensation" made by the payer nations comes to around USD 570 billion. As discussed under the methodology subsection, the

[4] The detailed methodology for the estimation of expenditure for green, renewable energy for the global economy and the various assumptions made to arrive at this number have been noted in Appendix 2 of Pollin (2015).

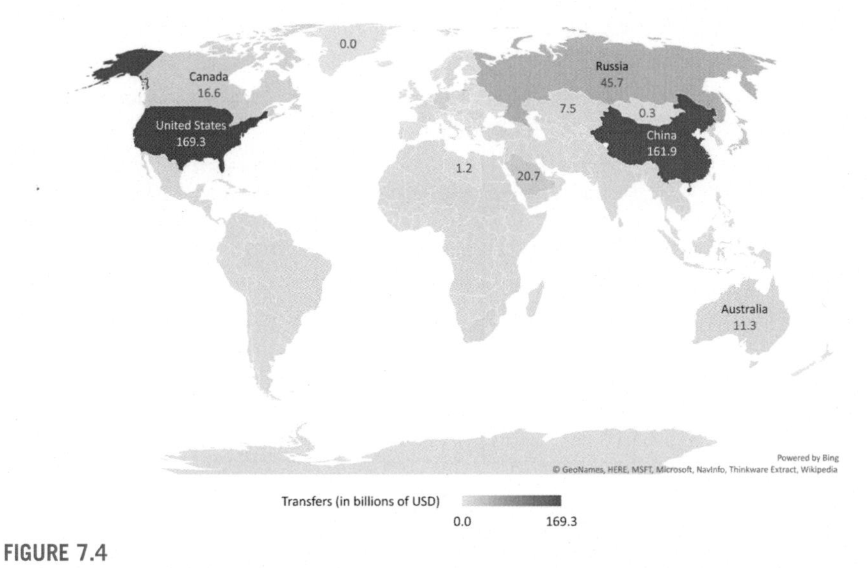

FIGURE 7.4

The annual transfers made by the "payer" countries (in billions of USD).

Source: Authors' calculations based on World Development Indicators, World Bank.

distribution of this amount across the payer countries is based on their distance from the global average (controlled for their population size). Using this formula, Fig. 7.4 shows the amount that these countries would have to contribute toward this global green deal fund. A few countries stand out, and it may help understand this policy if we discuss them. The two top countries in terms of absolute amounts of transfers are the United States and China. While it is not surprising to find the United States here because of its high per capita emissions, China needs some explanation. The reason why China is in this list to begin with is because its current per capita emission (7.54 metric tons of CO_2) is higher than the global average, but the reason why in absolute amount of compensation it is next only to the United States is because of its population size. By making the compensation amount a function of both per capita emission and population size, we have struck a balance between the two warring worlds of the global North and South on mitigating strategies. We believe it will be incorrect to err on either side completely as most of the current debates on climate change have been.

The other side of the same coin are the beneficiary countries that receive this compensation according to their distance below the global average controlled for their population size. Table 7.1 presents a bird's eye view of the countries getting the "green" compensation, and they are mostly underdeveloped when measured in terms of access to clean fuels and technology for cooking or electricity. In the absence of such a policy that gives them an opportunity to make this transition through funding from outside will help them break away from this vicious cycle.

Table 7.1 The countries receiving subsidy for moving to clean energy.

Access to energy	Access to clean fuels and technologies for cooking (% of population)		Access to electricity (% of population)	
	Number of countries	(In percentage)	Number of countries	(In percentage)
Below 50%	63	48.8	35	26.1
Above 50% to below 75%	18	14.0	17	12.7
Above 75% to below 90%	14	10.9	15	11.2
Above 90% to 100%	34	26.4	67	50.0
Total number of countries	129		134	

Notes: The total number of countries analyzed here is 134.
However, data for access to clean energy are available for 129 countries only.
Source: Authors' Calculations based on World Development Indicators, World Bank.

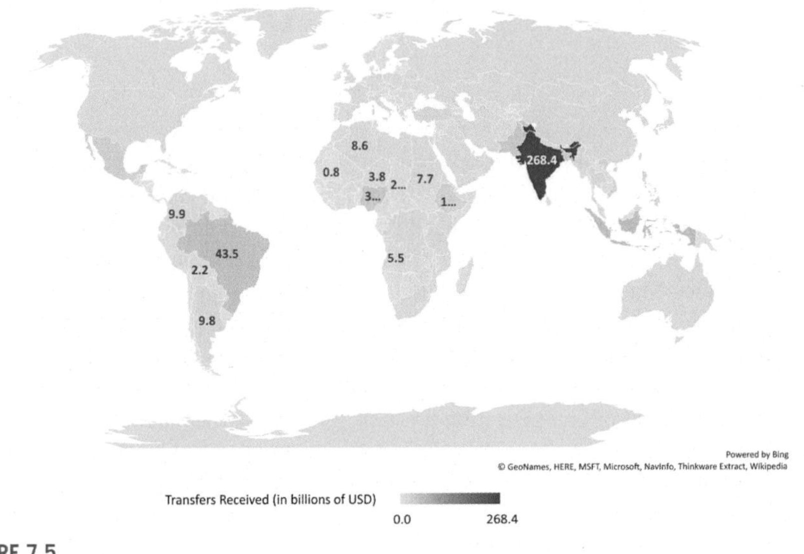

FIGURE 7.5

The annual transfers received by the "beneficiary" countries (in billions of USD).

Source: Authors' calculations based on World Development Indicators, World Bank Data.

A more detailed picture that arises from this is given by Fig. 7.5. A few cases stand out. The top beneficiary (in absolute compensation) in this turns out to be India, which is not surprising because of both its population size and its distance from the global emissions' average (India has per capita emissions of 1.73 metric ton as opposed to the global average of 4.97). The other suspects are all countries from the global South, but this list too springs a few surprises like France, Sweden, and Switzerland. What this tells us is that even high-income countries that have kept their per capita emissions low are beneficiaries of this globally just policy. With China in the first list and these countries in the second, it is obvious what this policy wants to achieve. It wants the countries to climb the emissions ladder below the global average without necessarily having to give up on their standard of living.

With the above discussion, one could get the impression that this policy is being a little too harsh on the emerging market economies such as China even as certain high-income countries end up being compensated in monetary terms. It would, however, only be half the picture. It is quite possible, as is indeed the case, that despite the compensation that China has to make, its carbon tax rates are lower than the global tax rates (Fig. 7.6). So, in that sense, the burden of adjustment is only partially falling on their shoulder and only to the extent that they emit more than the global average. Once they bring that under control, either through changes in the consumption pattern or energy mix or both, they will not have to carry the cross for the world.

In a similar manner, we can calculate the effective tax rate for the beneficiary countries (Fig. 7.7). This is especially important because the political opposition to any carbon tax is going to be the highest in these countries since they emit less than the global average and yet would have had to face the same carbon tax in the absence of this policy. What matters after this carbon compensation is the *effective* tax rate, i.e., the tax rate after deducting the transfer received out of the global green deal fund.

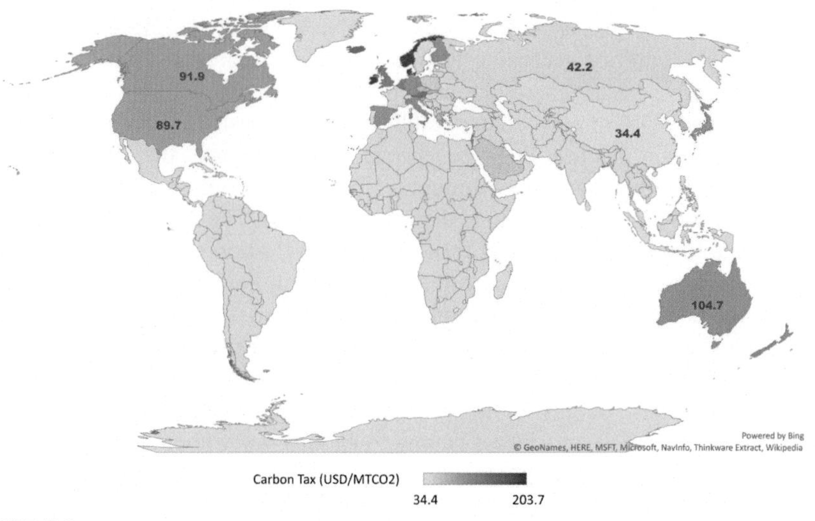

FIGURE 7.6

Carbon tax in the "payer" countries—a few selected countries.

Source: Authors' calculations. See text for details.

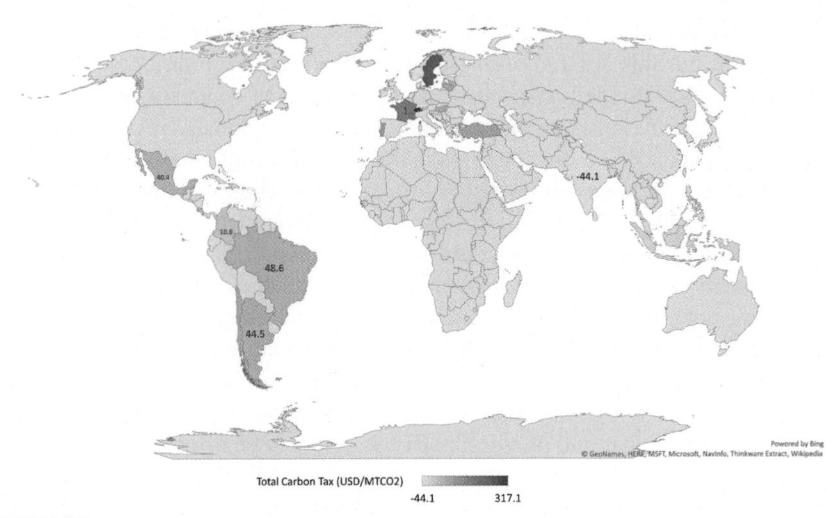

FIGURE 7.7

Carbon tax in the "beneficiary" countries—a few selected countries.

Source: Author' calculations. See text for details.

Conclusion

This study makes a case for a differential carbon tax across the globe so that this unjust distribution of burden between the South and the North is addressed since a large part of the stock of emissions have been contributed to by the global North in the past. A historically just approach would involve a global sharing of the burden between these countries according to their respective shares in global emissions. A comprehensive green energy transition for the world requires, according to most estimates, 1.5% of each country's GDP. Our proposal, JET, argues that countries who are emitting less than the current global average, the "beneficiaries," receive a part of their resources through carbon taxes from other countries, "the payers" who are emitting more than the global average. The latter will consequentially have to mobilize more taxes than their own transition requirements. Most of the climate change negotiations continue to deal with how much of the burden is to be shared between the North and the South. In the absence of a common agreement on this, environment has become the casualty. Our proposal, in contrast, will not only be a historically just way to balance the environmental and ecological damages done in the past, but it will also help the resource-poor developing countries to make the energy transition without having to unduly worry about the finances.

However, a few caveats apply to this study. Firstly, we have assumed that 1.5% of annual GDP is required for the next 20 years for all countries. This assumption may be far-fetched as some countries might have to spend much less than that, and some may be more. It is best to decide each case based on the individual country's energy requirements and their current levels of investment in green energy program.

Secondly, within these countries, it is better to do a class-wise analysis to find out who will bear the burden of these proposed carbon taxes. It might be the case that since carbon tax is regressive, the burden falls disproportionately more on the poorer people within these countries. However, that is not our intention. Given the increasing inequality within the economies, it is best to formulate taxes within these individual economies based on their consumption pattern across the classes and also to formulate a policy mechanism by which the poor in these countries can be shielded from the added burden of this tax. Thirdly, although the investment projections for the green energy program are based on 20-year long-run data, the global per capita emission levels are current and static. However, it is beyond doubt that per capita emission levels will decrease globally across time if such a strategy of clean energy program is adopted. Hence, there might be a need for a dynamic tax structure, which will be in proportion to the decreasing level of emissions so as to mobilize resources for the energy transition, and simultaneously incentivize people to move away from the dirty fossil fuels. Fourthly, a historically just approach would involve taking into consideration the *stocks* of emissions made by the advanced countries, especially the European nations and the USA. However, in this study, we have made our calculations based only on *flows*. However, addressing all these questions is beyond the scope of the current research and might be of interest to future researchers.

References

Azad, R., Chakraborty, S., December 2018. Green Growth and the Right to Energy in India. Working Paper 477. Political Economy of Research Institute, JNU and UMASS-Amherst.

Boyce, J., Riddle, M., 2007. Cap and Dividend: How to Curb Global Warming while Protecting the Incomes of American Families. Technical Report. Political Economy Research Institute, University of Massachusetts at Amherst.

Boyce, J.K., 2018. Carbon pricing: effectiveness and equity. Ecological Economics 150, 52−61. https://doi.org/10.1016/j.ecolecon.2018.03.030.

Fremstad, A., Paul, M., May 2017. A Distributional Analysis of a Carbon Tax and Dividend in the United States. Working Paper 434. Political Economy Research Institute.

Martin, R., June 2015. Climate Change: Why the Tropical Poor Will Suffer Most. MIT Technology Review.

Mendelsohn, R., Dinar, A., Williams, L., 2006. The distributional impact of climate change on rich and poor countries. Environment and Development Economics 11, 159−178.

Pollin, R., 2015. Greening the Global Economy. The MIT Press.

Pollin, R., Chakraborty, S., 2015. An egalitarian green growth programme for India. Economic and Political Weekly 1 (42), 38−52.

Pollin, R., Garrett-Peltier, H., Heintz, J., Chakraborty, S., 2015. Global Green Growth: Clean Energy Industrial Investments and Expanding Job Opportunities. United Nations Industrial Development Organization and Global Green Growth Institute, Vienna and Seoul.

Steffen, W., Rockström, J., Richardson, K., Lenton, T.M., Folke, C., Liverman, D., Summerhayes, C.P., Barnosky, A.D., Cornell, S.E., Crucifix, M., Donges, J.F.,

Fetzer, I., Lade, S.J., Scheffer, M., Winkelmann, R., Schellnhuber, H.J., 2018. Trajectories of the earth system in the anthropocene. ISSN 0027-8424 Proceedings of the National Academy of Sciences 115 (33), 8252–8259. https://doi.org/10.1073/pnas.1810141115. URL. http://www.pnas.org/content/115/33/8252.

CHAPTER

Riders on the storm: how hard did Robert Gordon's environmental headwind blow in the past?

8

Magnus Lindmark, Sevil Acar

Introduction

Robert Gordon has in several articles and, more recently, in his in book on US growth argued that American households have experienced stagnating levels of well-being since the early 1970s (Gordon, 2000, 2010, 2012, 2016). Most aspects of ordinary, everyday life did, to Gordon, very little progress after the great innovations of the late 19th century up to the 1930s had been refined, improved, and diffused during the post—war era. American well-being has therefore, Gordon argues, an S-formed shape. Exactly how to measure the development of well-being has certainly been around for some time, and Gordon himself points at a number of well-known shortcomings of the gross domestic product (GDP) measure, including the huge problems associated with measuring qualitative improvement of consumer goods.

The present-day situation is furthermore distinguished by a number of headwinds that all contribute to higher costs, lower growth, and ultimately a lost ability to come up with new innovations that could match those of the past. One of these headwinds is the environment, or rather costs for environmental protection, which Gordon believes will cause higher future energy prices. If so, this headwind is not only blowing at the American economy, but on the whole world. Gordon is not fully explicit on the nature of the environmental headwind, but one can easily point at a few plausible mechanisms or principles discussed in environmental economics in which environmental problems may contribute to stagnating welfare. For one, environmental damage may manifest itself as economic costs as various traditional assets held by households, companies, and the public sector degrade at a faster rate than anticipated through normal depreciation. For a perspective of well-being, such damage may also include natural resources, which are included in the traditional national accounting system. This category would include subsoil assets, growing forests in timber tracts, and so forth. Well-being could also be affected by deterioration of assets outside the

Handbook of Green Economics. https://doi.org/10.1016/B978-0-12-816635-2.00008-0

traditional asset boundary, for instance by deterioration of human health and loss of esthetic and existential values of landscapes.

A second type of mechanism through which the headwind may affect the real economy and material well-being is through mitigation costs. Potential future environmental damage may call for preventive actions, including the development of clean technologies and investments in various forms of abatement equipment, which usually takes place at high costs, tending to drain the economy of other urgent investments. Still, various forms of environmental costs have been around in the American economy for a long period of time, as a vast literature on American environmental history has demonstrated.

This chapter explores plausible environmental effects on American well-being in a historical perspective, using quantitative data and a methodological approach, which draws from approaches used by environmental economists. This adds one dimension of welfare to Gordon's discussion with some possible outcomes of relevance for the historical interpretation of American growth. First, it may be hypothesized that the levels of environmental damage rose especially during the prosperous decades following the World War II (WW2), including the spread of motor vehicles, diffusion of air traffic, and increased energy consumption. If so, the traditional way of measuring economic progress, GDP, would exaggerate the true development of well-being. On the contrary, second, the true progress of the period after 1970 may have been underestimated if environmental damage actually decreased as a consequence of an environmental awakening among producers, consumers, and agents creating modern environmental policy.

Background and summary of the relevant literature

On well-being

Robert Gordon has argued that American well-being did not improve significantly after the 1970s as the great innovations of the late 19th century up to the 1930s lost their power to generate total factor productivity (TFP) growth. Information and communication technology has furthermore not been able to match these innovations of the past in terms of propelling economic growth. Gordon is even more concerned about the well-being of future generations due to several headwinds: "the combined influence of globalization, global warming, and pollution constitutes an important additional drag on future growth going well beyond the post-1970 slowdown in the impact of innovation." Climate change is likely to lead to extreme weather events that will decelerate future economic growth and increase insurance premiums. Besides, Gordon foresees that future carbon taxes and direct regulatory interventions (e.g., fuel economy standards) will channel investments toward R&D on energy efficiency, believing that such an orientation of investments will not contribute to economic growth as much as the innovations of the past.

This chapter aims to shed light on the US well-being from an environmental perspective and to evaluate Gordon's ideas as far as the environmental headwind is concerned. For this, we use alternative indicators of well-being rather than the traditional measures of economic growth or GDP-related measures. One such indicator is based on an environmentally adjusted version of economic production. This indicator draws on Weitzman (1976), who pointed out that the maximum attainable level of consumption that could be maintained forever without running down capital stocks should be defined as net national product (NNP). This is to say that NNP is the stationary equivalent of future consumption. In this sense, the total capital stock should not change with natural resource depletion or depreciation, yielding the equality of total current resource rents to current net investment in reproducible capital. In relation to Weitzman's environmentally adjusted NNP, Solow (1986) handles Hartwick's rule as a way of maintaining the capital stock intact to keep real consumption constant over time. Green NNP or environmentally adjusted NNP is obtained by further revision of NNP by deducting the value of natural capital depreciation and pollution damage and, in some cases, accounting for nonmarket amenities such as parks, landscape, nature, and recreational access (green NNP = NNP − natural capital depreciation − pollution damage + nonmarket amenities).

Another such adjusted measure is genuine saving (GS), which is simply the difference between NNP and consumption. GS suggests a substitution between different types of capital assets. If a decline in one of the capital assets is compensated by an increase in another asset, GS would not decline and sustainability would be ensured. For instance, the decrease in natural resource stocks could be offset if the resource rents are invested in human capital or physical capital. Investing all the rents from natural resources in the accumulation of physical and human capital can be a way to keep GS nonnegative and instantaneous utility constant and nondeclining. Kunnas et al. (2014) investigate the welfare effects of carbon dioxide emissions from a historical perspective for the US well-being from 1870 and onwards and find that carbon emissions, regardless of the carbon price used, have little influence on the overall trend of sustainable development measured by GS in the United States. On the other hand, they find striking results for more recent years toward the 2000s. Their use of a high price for carbon ($312 per ton of carbon in 2001 in 2000 prices) from the BAU scenario in the Stern Review substantially drops GS in the early 1980s and even makes it negative for 1980, 1982, and 1983 in the United States. Augmenting the GS measure by incorporating an exogenous value of technological progress as proxied by changes in TFP in the United States, Hanley et al. (2016) find negative GS during the Great Depression and at the end of the WW2, whereas they detect positive GS in the rest of the post-1970 time frame. Through econometric analysis, the authors confirm that it was the Great Depression, which disrupted the stability of economic relationships between GS and the present value of changes in future consumption. Blum et al. (2017) similarly find that the US GS almost doubled after the WW2 (between 1950 and 1970), but the World War I (WW1) did not affect the US GS, whereas the GS of Great Britain, France, Australia, and Germany turned to negative during the WW1. Besides, in the United States as well as in Great Britain and Germany, technology is found to be the largest contributor to wealth accumulation during the second half of the 20th century. The authors point to the large differences in GS between Latin American and selected developed countries (Great Britain, Germany, Switzerland, France, the United States, and Australia) when TFP is included, which, as they state, could be a signal of a natural resource curse or technological gaps between these countries.

On valuation of environmental damages

In general, the method of valuation of environmental damages is a matter of translating emissions into ambient air concentrations and then into damage to human health, physical capital, agriculture, plants and animals, land use, and so on. Then, physical effects are turned into economic values using unit social costs of emissions. One problem about this task relates to the expected lifetime of pollutants or whether they are flow or stock emissions. Another problem is whether the pollutants have radioactive effects or can be absorbed by nature or transformed into harmless forms.

In the case of carbon emissions and other greenhouse gases (GHGs), one needs to specify a damage function that will translate the atmospheric concentration of GHGs into temperature changes, and temperature changes into damage caused by global warming. The importance of damage function specifications comes from the fact that different functions may give rise to different policy implications. However, in practice, it is extremely difficult to specify a damage function that will provide with the most useful and tractable results, as there are several structural uncertainties in climate science including the lack of knowledge on climate sensitivity and on how to represent damages from global warming (Weitzman, 2010). Integrated assessment models (IAMs) have been frequently used to assess the social cost of carbon (SCC). The purpose of such models is to represent dynamic interaction between environmental impact and the economy. They use different arbitrarily chosen damage functions (for CO_2). IPCC TAR Working Group II (2001) exemplifies the following alternatives among many others that are illustrated in IAMs:

1. Linear function: $D_t = f(T_t)$
2. Quadratic function: $D_t = f(T_t^2)$
3. Cubic function: $D_t = f(T_t^3)$
4. Hockey stick function, which assumes that impacts are approximately proportional to temperature change until a critical threshold is reached, implying rapidly worsening impacts after that threshold
5. Exponential function: $D_t = e^{f(T)}_t$

where D is damage, t is time, and T is the global mean temperature in deviation from the preindustrial times. Among them, the linear damage function assumes that impacts are proportional to temperature change since preindustrial times; the quadratic function assumes that impacts are proportional to a change in temperature to the power of two; and the cubic function assumes that impacts are proportional to temperature change to the power of three (IPCC 2001). Stern (2006), Nordhaus (2008) and Espagne et al. (2018) use quadratic damage functions in their PAGE, DICE, and RESPONSE models, respectively. Nordhaus (2007) finds that there will be a 0.03% GDP loss from the agricultural sector and 0.88% GDP decline from the nonagricultural sectors as a result of a 2.5°C warming for the United States. Following Nordhaus (2007) and specifying quadratic damage functions of both agricultural and nonagricultural sectors, Engström (2016) runs a multisector growth model where emissions from fossil fuels lead to environmental externalities. He finds slightly above 6% of GDP loss after 200 years (starting from 2005) for the nonagricultural sector, whereas the agricultural damages never climb above 1% of GDP.

Insley et al (2018), on the other hand, recommend an exponential damage function since the application of a quadratic function led to very little emission mitigation even at very high temperatures and carbon stock levels. In their model, the use of an exponential function implies that "damages from climate change would be disastrous," for temperature increases above 3°C (Insley et al 2018, p. 25).

Peck and Teisberg (1994) utilize linear and cubic form damage functions. IPCC TAR Working Group II (2001) states that "*a cubic function implies low near-term impacts but rapidly increasing impacts further in the future. Using conventional discounting, this means that early emissions under a cubic damage function will cause less damage over their atmospheric lifetime, compared to a scenario with linear damages. The marginal damage caused by emissions further in the future, on the other hand, is much higher if we assume a cubic damage function*" (pg. 944). When utilizing a linear damage function in temperature increase, Peck and Teisberg (1994) find that optimal carbon tax remains below 35 per ton until the year 2200, and needless to say, with such a tax, mitigation of CO_2 emissions in comparison with the case where there is no emission control remains very limited. On the other hand, when a cubic function is assumed, "marginal warming costs are much greater than with the linear function, once the temperature level exceeds 1.73°C" (Peck and Teisberg, 1994, p. 295). In that scenario, the optimal carbon tax starts at around $10 per ton and then jumps to $208 per ton by early 2100s. This implies that such a high carbon tax will be appropriate to combat emissions by employing carbon-free technologies. What also matters according to the study is whether warming costs are related to the level of warming or to the rate of warming.

In summary, various authors (e.g., Peck and Teisberg, 1994; Kandlikar, 1996; Tol, 1996; Kosugi et al., 2009; Anda et al., 2009; Weitzman, 2010; Ekholm et al., 2013; Stern, 2013) confirm that the functional form of the selected damage function has strong implications for the value of the damage for the emission in consideration, although the choice of this function is always arbitrary.

What matters a lot for carbon damage calculation apart from the choice of the damage function is the issue of discounting. Since SCC is time dependent (the damage is not immediate but occurs in the future), historical price of carbon needs to be adjusted by calculating the historical discounted unit cost based on a chosen interest rate. The choice of the discount rate is a widely discussed problem in environmental accounting. Tol (2008) conducts a metaanalysis for the 2011 estimates of the SCC, and he evidences that a lower discount rate implies a higher estimate. The issue becomes highly apparent in the comparison of Stern (2006) and Nordhaus (2008) estimates. Stern finds alarming results for carbon costs possibly driven by a relatively low interest rate (1.4%), whereas Nordhaus uses a 4.1% interest rate as a result of the so-called Ramsey rule below:

$$r = \rho + \eta g$$

where g is the future growth of consumption; η is a parameter that describes the rate of falling marginal utility of consumption; and r is the discount rate (interest rate). ρ is the pure time preference. Stern chooses a low rate of time preference (0.1%) in his model, whereas Nordhaus uses 1.5% rate of pure time preference. Although both authors use dynamic IAMs and isoelastic social utility functions in their analyses, and both assume an annual economic growth rate of 1.3% over the next century, they end up with contrasting results, most probably stemming from their differing discount rate choices.

Notice that Gordon believes that g may be very low in the future. He even thinks it may return to preindustrial revolution levels (around 0.7%). He explains the low g through basically other forces than the environment.

We continue with alternative calculations of a green NNP for the US accounting for the social costs of three selected pollutants in the rest of the chapter.

General approach

The purpose of this study is to explore how the "environmental headwind," historically and in the present, has affected the traditional measure of economic progress, GDP per capita in the United States from 1870 to 2010. The headwind is approximated as the social cost of three pollutants, of which two, sulfur dioxide (SO_2) and nitrogen oxides (NO_x), represent "traditional" forms of environmental damage, while the third, carbon dioxide (CO_2), represents global warming. To estimate the impact on economic welfare, we deduct these environmental costs from GDP.

We are using contemporary estimates of environmental costs to estimate the historical development of environmental damages of the abovementioned pollutants for which reasonable data exist. Due to basic uncertainty concerning what is known as a "damage function," we will elaborate a few alternative measures of these costs. In a second step, we use these costs to adjust the US NNP from 1950 and onward. This results in an environmentally adjusted NNP, sometimes known as the green NNP, as previously elaborated in Section 2. Finally, the traditional GDP and the green NNP measures are compared and analyzed. This approach was previously used in Lindmark and Acar (2013, 2014) and Kunnas et al. (2014). The analytical approach is to compare whether the growth rates differ over the period in question. A few basic results can be foreseen. For scenario one, it is possible that GDP and the green NNP have developed in a similar way. This would, for instance, be the case if the social cost of emissions has been small compared with GDP. The implication would in that case be that the environmental headwind has not affected the general development of welfare in any significant way. In scenario 2, the environmental headwind may have caused the green NNP to develop at a significantly lower rate than GDP during the 1950s and 1960s, but at a higher rate after 1970 as environmental policy and technical progress caused the rate of emissions to decrease. This means that, in scenario 2, GDP has over- and underestimated the true progress of welfare over certain subperiods, which implies that Gordon's view on performance of the US economy should be adjusted. Finally, in scenario 3, green NNP has only recently shown a slower growth rate as compared with GDP, implying that Gordon is probably right in claiming that environmental costs are likely to lower the future development of welfare and that environmental degradation is indeed a significant headwind for the future to come.

Data and methodological issues

The sulfur data used in this study have been collected from Lefohn et al. (1999) for the period 1870–1970. These data are presented as benchmarks every 10th year. That is why we have used a straight linear interpolation between the benchmarks. The data from 1970 to 2010 were collected from the European Commission Joint Research Centre. NO_X emissions data between 1900 and 1970 have been collected from EPA (1996). These data are also estimated for every 10th year. Also, NO_X data from 1970 to 2010 have been collected from the European Commission Joint Research Centre. Carbon emissions data have been collected from Boden et al. (2016) and cover emissions from combustion of fossil fuels and cement production.

Social costs of emissions have been collected from a study by Jaramillo and Muller (2016) focusing on the US energy sector. Jaramillo and Muller's study uses an IAM. The specific model used in their study connects emissions to monetary damages using six modules: emissions, air quality modeling, concentration, exposure, dose-response, and valuation. In our study, we base the historical cost estimates on the contemporary marginal damage costs for 2011 as reported by Jaramillo and Muller (2016). Thus, the original emissions series, expressed in physical units, have been transformed to economic volumes expressed in 2011 damage costs. By doing so, we follow previous historical investigations of the social cost of environmental impact, such as Lindmark and Acar (2013) and Kunnas et al. (2014). Jaramillo and Muller find that total damages across oil and coal extraction, oil refineries, and electricity production amounted to $175 billion (in 2000$) in 2002. In 2011, air pollution damages associated with emissions from these sectors totaled $131 billion (in 2000$). For enabling comparisons with previous studies, Jaramillo and Muller further adjusted their cost estimate, using their 2011 marginal damages. The unit social costs used in our investigation are $14,000 per ton for SO_2 and $4600 for NO_X.

There have been several attempts to estimate the SCC since the 1990s. IAMs are also in this case the basic methodological tool. The most widely used are probably the DICE, RICE, MERGE, and PAGE models. There are, however, some principal differences as compared with most other emissions, which may be worth noticing. First, the damages from GHGs are expected to occur in the future. This means that any cost estimate will need to make assumptions on or make a choice over an appropriate interest rate. Certainly, this is in principle true for other emissions, but as these emissions cause damages in the near future, the impact of the interest rate will be smaller. There are also several studies, which demonstrate that the choice of interest rate is a more important factor than the choice of IAM as such. For instance, Nordhaus used a discount rate at 1.5% per year, using market interest rates as a guide, and found that the optimal tax, which equals the marginal damage cost, ought to be roughly $30 per ton of carbon (Nordhaus and Boyer, 2000). Stern (2006), in contrast, used a discount rate of 0.1%, arguing that climate change puts an additional moral obligation on present generations to consider the well-being of future generations, and concluded that a tax of $250 per ton of carbon is optimal. As interest rates are so important in the case of GHG emissions, Lindmark and Acar (2014) made a historical adjustment of the contemporary cost, so that it decreased historically at the same rate with the interest rate used in the specific IAM. This means that the social unit cost decreased exponentially in the past, reflecting the time effect of past generations. We further notice that the contemporary choice of the interest rate, by which the cost of future global warming should be discounted, also affects the historical environmental costs that are estimated. The view on the future therefore also determines the historical narrative.

Another aspect of the SCC is the risk of catastrophic effects, a problem pointed out by Weitzman (2009). In short, his argument was that even a small risk of truly catastrophic developments because of global warming represents a fundamental problem when estimating the SCC. Even a very high interest rate will be offset by a tiny risk of global warming causing the destruction of humankind, implying infinite damage costs. To deal with this problem, known as "Weitzman's dismal theorem," valuation can be approached as an insurance-like problem, rather than as a cost-benefit issue.

Golosov et al. (2014) presented a new model with a better representation of natural scientific aspects of global warming and recalculated Nordhaus' and Stern's estimates of the SCC arriving at costs at a little under $60 per ton of carbon when using Nordhaus' 1.5% discount rate and around $500 per ton when using Stern's 0.1% rate. Golosov et al. also presented estimates of both Nordhaus' and Stern's SCC given the risk of catastrophic events, arriving at $489 per ton for Nordhaus and $4263 per ton for Stern. The latter cost is so high that it reduces all other components of the analyses. It will therefore not be used in the study.

Notice that a low discount rate, as assumed in Stern's estimate of the cost of carbon, leads to high unit SCC in the present but will also have implications for the historical estimate such that the same interest rate should be assumed also for past generations. After all, neither Stern nor Nordhaus knows the "true" discount rate in the present, which implies an ignorance they must have shared also with past generations. However, if we use a fixed 2010 SCC for monetarization of the historical series, we will run into methodological problems. This is so since a fixed SCC implies that past generations must have discounted the future at a low rate, which then gradually increased. This is the only way to ensure the same unit price over time. A low but increasing discount rate implies in turn that past generations initially valued the well-being of future generations higher than we do today. This is certainly implausible. The factor that is changing is, however, the discounting period as the costs of global warming, in the perspective of past generations, occur in a more distant future as it is the case today. The historical SCC should therefore be adjusted with time. Here we are using the basic formula that the discounted value equals the present value times the discount rate raised to the power of the number of time periods. The present value equals the contemporary carbon price, in the Nordhaus case $60 per ton of carbon dioxide in 2010. Thus, the SCC in 1870 was $60/(1015^140) or $7 per ton of carbon. Notice that the cost is expressed in the 2010 price level. In the case of Stern, the SCC in 1870 was $500/(1001^140) or $435. The implicit assumption is that past generations discounted the future at the same rate as present generations, or rather as two influential environmental economists, given that past generations had the same limited knowledge on the SCC as captured in contemporary IAMs. We also do the same calculation for Nordhaus' cost in the presence of catastrophic events. A similar exercise is not necessary for SO_2 and NO_X since the damages caused by these pollutants largely affect human health and these damages generally occur without any significant time delay.

Results

Historical development of emissions

Fig. 8.1 shows the development of the emissions in physical units over the full investigated period.

Sulfur emissions increased more rapidly than carbon and NO_X up until the early 1920s. After 1920, sulfur emissions stabilized and remained at this level until the late 1970s. From the late 1970s onward, there is evidence for a historically rapid decline of the sulfur emissions, and at the end of the period, they were down at approximately the same level as in the early 1890s.

Carbon emissions also experienced a rapid increase up to approximately 1920. While sulfur emissions stabilized, carbon emissions continued to increase even though the rate of increase was slower than previously. A second period of the rapid increase of carbon emissions occurred during the period 1960–1973. This period was followed by slightly falling emissions until the early 1980s, after which emissions increased at a comparatively low rate.

NO_X emissions followed a slightly higher growth rate than carbon emissions until the early 1970s, after which the NO_X emissions have fallen, however, at a slower rate than the sulfur emissions.

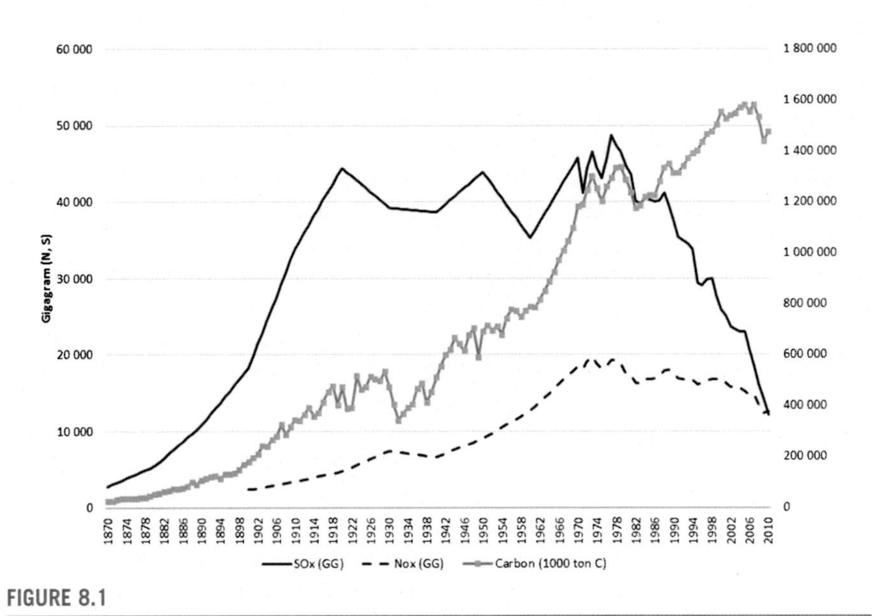

FIGURE 8.1

SO_2, NO_x, and carbon emissions in the United States, 1870–2010.

Source: Authors' Own Illustration.

These emissions therefore represent three different development paths. The different development path of sulfur as compared with carbon reflects that sulfur emissions to a high degree were process-related emissions from the copper industry, meaning that they arose in the production process itself and not only as a consequence of fuel combustion (Schmitz, 2000). The rapid development of sulfur emissions during the early part of the period was, for instance, related to the rapidly increasing demand for copper during the second industrial revolution (Uekötter, 2009). Electrical equipment, in Gordon's phrasing a key innovation of the past, such as wires, generators, and motors were all depending on copper. As the sulfur content of copper ores (copper sulfide) is comparatively high, the smelting process itself caused the release of sulfur with the flue gases. Copper smelters were therefore equipped with high chimneys for diluting the gases and spreading the sulfur downfall over a larger area. Still, the downfall of sulfur in the form of the so-called acid rain caused extensive damage on agricultural land and the death of cattle, which in turn led to lawsuits. Prime examples were the legal processes in Ducktown, Tennessee, which led to the US Supreme Court's first verdict in an air pollution case in 1907 (LeCain, 2009; Maysilles, 2011). An important political outcome was Theodore Roosevelt to establish the Anaconda Smelter Smoke Commission (1911–20), with the directive to suggest measures that the company should undertake to address the smoke issue. Another outcome was the development of electrical scrubbers, which were used to abate the sulfur emissions. The introduction of such technology, along with the legal framework envisaged by the court cases and the Anaconda

Smelter Smoke Commission, probably explains the stabilization of sulfur emissions over the period 1920–70.

Carbon emissions were, on the other hand, more or less directly a function of coal and oil combustion, with the exception of process emissions from cement production. These were, however, comparatively small as compared with the combustion of fossil fuels.

NO_X, finally, was mainly caused by chemical processes when air gets in contact with high-temperature metal surfaces in internal combustion engines, including jet engines. Internal combustion engines were, again, a basic innovation of the past which, according to Gordon, drove TFP growth and, thus, economic growth in the period from the 1920s to the 1970s. NO_X emissions were mainly abated through the introduction of catalytic converters in automobile exhaust pipes from approximately the mid-1970s.

Development of the environmentally adjusted net national product

Fig. 8.2 shows the development of the US GDP per capita and environmentally adjusted NNP in two versions, where the low estimate is based on the Golosov et al. (2014) carbon price based on Nordhaus' assumptions, and the high estimate

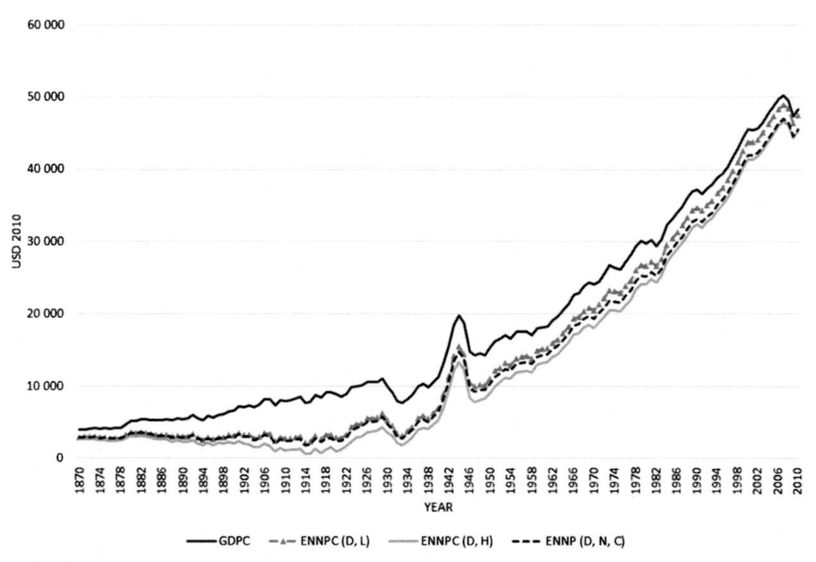

FIGURE 8.2

Environmentally adjusted NNP (high and low estimate) and GDP of the United States, 1870–2010 (in thousand 2010 USD).

Note: ENNPC (D, L) is the environmentally adjusted net national product using Nordhaus' 1.5% discount rate at $60/ton C in 2010 (low); ENNPC (D, H) is the environmentally adjusted net national product using Sterns' 0.1% discount rate at $500/ton C in 2010 (high); ENNPC (D, N, C) is the environmentally adjusted net national product using Nordhaus' 1.5% discount rate at $489/ton C in 2010.

Source: Authors' Own Illustration.

is using Stern's assumption. As seen from Fig. 8.2, both versions of the environmentally adjusted NNP are significantly lower than the traditional GDP.

The period up to approximately 1918 shows a nearly stagnating environmentally adjusted NNP in the low-bound estimate and even a negative development when the high-bound estimate is used. The high-bound environmentally adjusted NNP is even negative between 1907 and 1920. We will return shortly to the methodological implications of this.

The implications are in practice that Nordhaus' assumptions provide a comparatively low cost of carbon even today. Even if the higher interest rate (compared with Stern) gives the unit cost of carbon rises by a factor of 10, the total costs of carbon are not high enough to induce significant adjustment of the environmentally adjusted NNP. Stern's estimate, on the other hand, results in high costs in the present, but due to the low interest rate, the unit costs do not change very much. This means that the high-end estimate does not change very much neither. Still, the high-end estimate shows very low values around the 1920s, close to zero, which clearly seems implausible. This strongly suggests that the interest rate cannot have been as low as 0.1% in the past.

Fig. 8.3 shows the different versions of the environmentally adjusted NNP per capita relative to GDP per capita.

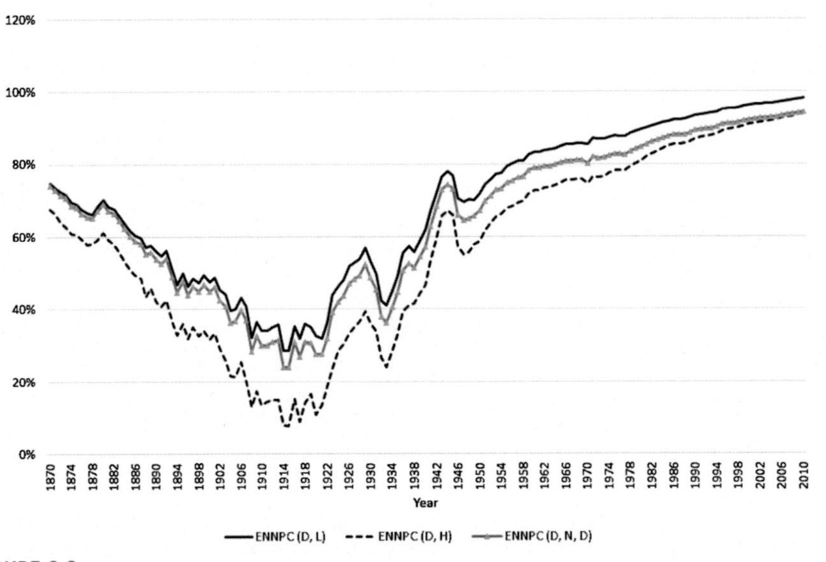

FIGURE 8.3

Environmentally adjusted NNP (high and low estimate) relative to GDP of the United States, 1870–2010 (in percentage).

Source: Authors' Own Illustration.

From Fig. 8.3, it is evident that all versions of the environmentally adjusted NNP developed at a lower rate than GDP during the period 1870–1910. This implies that the environmental headwind was indeed strong during the late 19th century, meaning that welfare adjusted for environmental effects developed at a lower rate as compared with income growth. In short, the great innovations of the 19th century appear as less great considering the environmental consequences due to foremost sulfur emissions. This characterization is also in line with the qualitative results from American environmental history (Uekötter, 2009; Rosen, 1995; McMillan, 2000). From c.1925 to 1955, all versions of the environmentally adjusted NNP increased at a faster rate than GDP, meaning that the environmental welfare increased at a faster rate than GDP. The environmentally adjusted NNP continued to grow faster from 1955 up to the present and, moreover, at a relatively constant rate. This means that the slowdown and stagnation in the American economy after 1970 as reported by Gordon is much less evident if environmental performance is taken into consideration. From the 1970s onward, there is also an obvious structural change in emissions as the costs of SO_2 and NO_x have decreased, whereas the cost of carbon has increased. This is shown in Fig. 8.4.

This increase is in turn the result of slowly increasing emissions in combination with the time effect. The shorter the discounting period becomes, the higher is the SCC. Therefore, as long as emissions remain high, the closer the US economy is moving toward materialization of the costs of global warming, possibly even

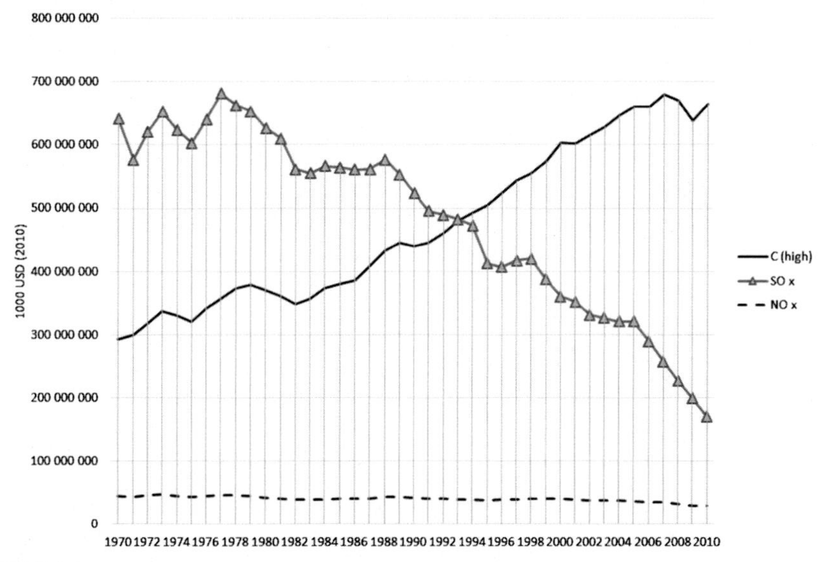

FIGURE 8.4

Social costs of carbon, SO_2 and NO_x in the United States, 1970–2010.

Source: Authors' Own Illustration.

experiencing catastrophic events. This means that Gordon is right in so much as the environmental headwind can be expected to increase in strength in the future to come.

Concluding remarks

This chapter has shown that the environmental headwinds of the American past were significant, especially during the period from the late 19th century up until the 1920s. However, the most important conclusion from our study is that carbon emissions and global warming represent a different type of problem as compared with the experiences of the past, where the principal difference is the time dimension. While traditional emissions had a more or less immediate impact, the impact of GHGs is more gradual and its consequences occur predominantly in the future. This means that GHGs cannot be treated in the same way as other pollutants. The SCC is simply rising over time, as a result of a pure time effect, meaning that carbon damage is rising as the world is moving toward sizable negative impacts of global warming. This is not the case for the other pollutants. We therefore introduce a price of carbon that is rising over time, capturing an essential aspect of global warming as a social and economic issue, rather than a purely scientific issue. The social and economic dimensions of carbon emissions are not captured if global warming is only indicated by the emissions of carbon measured in physical quantities, as the cost of carbon is higher today than yesterday, even if we adjust for inflation, since today is closer in time to the full realization of the costs of global warming than yesterday. Furthermore, the time effect, as it appears even at a low interest rate, captures a realistic view of global warming from a social perspective. In the past, the SCC had been low, even close to zero, reflecting a lowly perceived impact on well-being. In essence, global warming was an issue that only concerned a handful of scientists prior to the 1990s. Over time, however, the cost has risen exponentially, and global warming is today a larger problem than the traditional emissions. The environmental headwinds of the past were rising with the levels of emissions, whereas the headwind of global warming is additionally rising over time. Thus, time and a positive time preference is causing the headwind to blow harder and harder. A consequence of this will be both a rising impact on the future growth of well-being, as well as an increased transformation pressure to come up with the innovations of a greener future.

References

Anda, J., Golub, A., Strukova, E., 2009. Economics of climate change under uncertainty: benefits of flexibility. Energy Policy 37 (4), 1345–1355.

Blum, M., Ducoing, C., McLaughlin, E., 2017. A sustainable century? genuine savings in developing and developed countries, 1900–2000. In: Hamilton, K., Hepburn, C. (Eds.), National Wealth: What Is Missing, Why it Matters. Oxford University Press, Oxford, pp. 89–113.

Boden, T.A., Marland, G., Andres, R.J., 2015. Global, Regional, and National Fossil-Fuel CO_2 Emissions. Carbon Dioxide Information Analysis Center. Oak Ridge National Laboratory, U.S. Department of Energy, Oak Ridge, Tenn., U.S.A. https://doi.org/10.3334/CDIAC/00001_V2015.

Crippa, M., Janssens-Maenhout, G., Dentener, F., Guizzardi, D., Sindelarova, K., Muntean, M., Van Dingenen, R., Granie, C., 2016. Forty years of improvements in European air quality: regional policy-industry interactions with global impacts. Atmospheric Chemistry and Physics 16 (6), 3825–3841.

Ekholm, T., Lindroos, T.J., Savolainen, I., 2013. Robustness of climate metrics under climate policy ambiguity. Environmental Science & Policy 31, 44–52.

Engström, G., 2016. Structural and climatic change. Structural Change and Economic Dynamics 37, 62–74.

Espagne, E., Pottier, A., Fabert, B.P., Nadaud, F., Dumas, P., 2018. SCCs and the use of IAMs: let's separate the wheat from the chaff. International Economics, Elsevier 155 (C), 29–47.

Golosov, M., Hassler, J., Krusell, P., Tsyvinski, A., 2014. Optimal taxes on fossil fuel in general equilibrium. Econometrica 82 (1), 41–88.

Gordon, R.J., 2016. The Rise and Fall of American Growth: The U.S. Standard of Living since the Civil War. Princeton University Press.

Gordon, R.J., 2012. Is U.S. Economic Growth over? Faltering Innovation Confronts the Six Headwinds. NBER Working Paper No. 18315.

Gordon, R.J., 2010. Revisiting U.S. Productivity Growth over the Past Century with a View of the Future. NBER Working Paper 15834, March.

Gordon, R.J., 2000. Does the new economy measure up to the great inventions of the past? The Journal of Economic Perspectives 14 (Fall, 4), 49–74.

Hanley, N., Oxley, L., Greasley, D., McLaughlin, E., Blum, M., 2016. Empirical testing of genuine savings as an indicator of weak sustainability: a three-country analysis of long-run trends. Environmental and Resource Economics 63 (2), 313–338. https://doi.org/10.1007/s10640-015-9928-7.

Insley, M., Snoddon, T., Forsyth, P.A., 2018. Strategic interactions and uncertainty in decisions to curb greenhouse gas emissions. University of Waterloo, Department of Economics Working Papers, No: 1805. https://cs.uwaterloo.ca/~paforsyt/insley_snoddon_forsyth_working_paper_Dec_2017.pdf. (Accessed 5 July 2019).

IPCC TAR Working Group II, 2001. Impacts, adaptation and vulnerability. In: Climate Change. (Chapter 19: Vulnerability to Climate Change and Reasons for Concern: A Synthesis).

Jaramillo, P., Muller, N., 2016. Air pollution emissions and damages from energy production in the U.S.: 2002–2011. Energy Policy 90, 202–211.

Kandlikar, M., 1996. Indices for comparing greenhouse gas emissions: integrating science and economics. Energy Economics 18 (4), 265–281.

Kosugi, T., Tokimatsu, K., Kurosawa, A., Itsubo, N., Yagita, H., Sakagami, M., 2009. Internalization of the external costs of global environmental damage in an integrated assessment model. Energy Policy 37 (7), 2664–2678.

Kunnas, J., McLaughlin, E., Hanley, N., Greasley, D., Oxley, L., Warde, P., 2014. Counting carbon: historic emissions from fossil fuels, long-run measures of sustainable development and carbon debt. Scandinavian Economic History Review 62 (3), 243–265. https://doi.org/10.1080/03585522.2014.896284.

LeCain, T.J., 2009. Mass Destruction. The Men and Giant Mines that Wired America and Scarred the Planet. Rutgers University Press, Piscataway, N.J.

Lefohn, A.S., Husar, J.D., Husar, R.B., 1999. Estimating historical anthropogenic global sulphur emission patterns for the period 1850–1990. Atmospheric Environment 33 (21), 3435–3444.

Lindmark, M., Acar, S., 2014. The environmental Kuznets curve and the pasteur effect: environmental costs in Sweden 1850–2000. European Review of Economic History 18 (3), 306–323.

Lindmark, M., Acar, S., 2013. Sustainability in the making? A historical estimate of Swedish sustainable and unsustainable development 1850–2010. Ecological Economics 86, 176–187.

Maysilles, D., 2011. Ducktown Smoke: The Fight Over One of the South's Greatest Environmental Disasters. University of North Carolina Press, Chapel Hill.

McMillan, D., 2000. Smoke Wars: Anaconda Copper, Montana Air Pollution, and the Courts, 1890–1924. Montana Historical Society, Helena.

Nordhaus, W.D., Boyer, J., 2000. Warming the World Economic Models of Global Warming. MIT-Press, Cambridge, Massachusetts.

Nordhaus, W., 2007. A review of the stern review on the economics of climate change. Journal of Economic Literature 45 (3), 686–702.

Nordhaus, W., 2008. A Question of Balance – Weighing the Options on Global Warming Policies. Yale University Press.

Peck, S., Teisberg, T., 1994. Optimal carbon emissions trajectories when damages depend on the rate or level of global warming. Climatic Change 28 (3), 289–314.

Rosen, C.M., 1995. Businessmen against pollution in late nineteenth century Chicago. Business History Review 69 (3), 351–397.

Schmitz, C.J., 2000. The world copper industry: geology, mining techniques and corporate growth 1870–1939. Journal of European Economic History 29 (1), 77–105, 77-78.

Solow, R.M., 1986. On the intergenerational allocation of natural resources. The Scandinavian Journal of Economics 88 (1), 141–149.

Stern, N.H., Peters, S., Bakhshi, V., Bowen, A., Cameron, C., Catovsky, S., Crane, D., Cruickshank, S., Dietz, S., Edmonson, N., Garbett, S.-L., Hamid, L., Hoffman, G., Ingram, D., Jones, B., Patmore, N., Radcliffe, H., Sathiyarajah, R., Stock, M., Taylor, C., Vernon, T., Wanjie, H., Zenghelis, D., 2006. Stern Review: The Economics of Climate Change. Cambridge University Press, Cambridge.

Stern, N., 2013. The structure of economic modeling of the potential impacts of climate change: grafting gross underestimation of risk onto already narrow science models. Journal of Economic Literature 51 (3), 838–859.

Tol, R.S.J., 1996. The damage costs of climate change towards a dynamic representation. Ecological Economics 19 (1), 67–90.

Tol, R.S.J., 2008. "The social cost of carbon: trends, outliers and catastrophes", economics—the open-access. Open-Assessment E-Journal 2 (25), 1–24.

Uekötter, F., 2009. The Age of Smoke: Environmental Policy in Germany and the United States, 1880–1970. University of Pittsburgh Press.

Weitzman, M.L., 2009. On Modeling and Interpreting the Economics of Catastrophic Climate Change. Review of Economics and Statistics 91 (1), 1–19.

Weitzman, M.L., 2010. What is the "damages function" for global warming – and what difference might it make? Climate Change Economics [Internet] 1 (1), 57–69.

Weitzman, M.L., 1976. On the welfare significance of national product in a dynamic economy. Quarterly Journal of Economics 90 (1), 156–162.

Further reading

Hamilton, K., Atkinson, G., Pearce, D., 1997. Genuine savings as an indicator of sustainability. CSERGE Working Papers, GEC 97-03, pp. 1–28.
Hamilton, K., Clemens, M., 1999. Genuine savings rates in developing countries. The World Bank Economic Review 13 (2), 333–356.
Mäler, K.G., 1991. National accounts and environmental resources. Environmental and Resource Economics 1 (1), 1–15.
Nordhaus, W., 1991. To slow or not to slow: the economics of the greenhouse effect. Economic Journal 101 (407), 920–937.

CHAPTER

Economic instruments of greening

9

A. Erinç Yeldan

Introduction

This chapter is about instruments. What are the tools of policy interventions that economists offer in controlling environmental pollution within a market system? What are the advantages and disadvantages of each? And, what kind of an institutional infrastructure is needed to enforce such interventions?

In very broad terms, the arsenal of effective abatement policies can be categorized under three modalities: (1) imposition of a price for the pollutant that would reflect not only the private but also all the social costs of its impact; (2) fiscal policies and subsidies toward utilization of more eco-friendly modes of production; and (3) implementation of standards and regulations that would lead the economic agents (consumers, producers, investors, etc.) to affect their decisions toward more environmentally friendly outcomes. The two general types of instruments that the authorities could install under a market system, in turn, are either based on *pricing* or *quantity setting* within the so-called *emission trading scheme* (ETS). Both instruments have desirable and undesirable side effects, as no policy intervention is 100% free of risks of policy distortion. The art of economic interventions is to balance against the costs and gains of the interventions to the market system.

In dealing with any public good such as maintaining environmental affluence, the key problem is markets' failure in delivering a *price* on the quality of environment per se that would *internalize* the social costs against its degradation. If markets could have generated a mechanism that would correctly reflect all the social costs of production including pollution and its effects on human welfare, then it would have been a trivial task, and there would be no need for a regulatory body imposing restrictions to unfettered workings of the market. The origin of the matter, however, does not admit such a possibility, and regulatory interventions become generally necessary to be able to *internalize* the otherwise unaccounted costs within the free market calculations of private optimum.

To make an effective assessment of what is involved, it would be proper to first work with an abstract example to illustrate some of the common themes and key concepts. This will be the topic of the next section of this chapter. In what follows, we will look at the sets of policy instruments under Sections 3 and 4; in Section 5, we

Handbook of Green Economics. https://doi.org/10.1016/B978-0-12-816635-2.00009-2

will focus on the realities of the global economy that surround such decisions and conclude with an overall assessment.

Theorizing pollution abatement

We first start with conceptualization of the problem at hand. Let us remember that effectiveness of policy instrument in any field of economics relies on one main principle: *maximization of the benefits* involved. This is equivalent, under certain general assumptions that we do not have room to further elaborate, to *minimization of costs*—on environment, on agents' budgets, on producers' profit opportunities, on government's budgetary outlays, etc. In our case, we can distinguish two types of costs: (1) the very cost of environmental pollution. We will capture these costs within the concept of "*marginal environmental damage cost*" (MEDC). This is the increase in total environmental cost due to a unit increase in pollution. (2) Costs of abatement. Expressed as the increase in aggregate abatement costs due to a unit of environmental pollution avoided, we get the *marginal abatement cost* (MAC). Typically, MEDC increases as more and more environmental pollution is inflicted, and likewise, MAC is upward sloping against the amount of pollution prevented.

Theory suggests that maximum social efficiency is attained when these two types of marginal costs are equalized against the quantity of pollution generated. To illustrate this, consider the following diagram where marginal costs (in, say, US$) are plotted against aggregate quantity of pollution generated (in tons).

The MEDC curve slopes upward as noted: as quantity of pollution increases, the marginal damage cost on the environment also increases exponentially. MAC is *downward* sloping since as we let go pollution, our costs of abatement decline; if otherwise we pursue to *reduce* the amount of pollution generated (we move toward back to the origin along the horizontal axis), it becomes costlier to impose additional restrictions to control for pollution; hence marginal costs of abatement tend to rise as we approach back to the origin.

At Q*, we reach an optimum in the sense that the environmental damage caused by a marginal unit of pollution is equated with the *marginal benefit* (inverse of the marginal cost of abatement) obtained from its reduction. To the right of Q*, *too* much pollution is allowed for, and reducing pollution toward Q* from right to left, marginal benefits from abatement exceed marginal costs of damage that had been inflicted. Thus, reducing pollution improves social welfare. A more controversial action regards the levels of *excessive* pollution control. To the left of Q*, marginal costs of abatement *exceed* (the marginal benefits of more environmental protection fall short of) the marginal damage costs on the environment, thus cutting back excessive regulation and allowing *more pollution*; thereby moving back to Q* is socially welfare enhancing. This is to say that *optimal level of pollution is not necessarily zero*. For smooth functioning of the economy, given the levels of technological advance, institutional infrastructure, consumer preferences, and so on, a positive level of environmental pollution is unavoidable. What Fig. 9.1 narrates is the *optimal* level of such pollution.

Now that we agreed upon the optimal level of pollution tolerable, the next issue in line is how to *impose* it. Thereby, suppose we regard Q* as the aggregate (optimal) carbon budget for our purposes, and we need to allocate this magnitude among its generators. A complete detail of the issues to be addressed in this endeavor is surely beyond the scope of this chapter; yet, we will illustrate the main principles that will guide us to the solution to this question with the aid of a graphical explanation covering the decisions of two competitive firms (or national economies).

Suppose two firms A and B are given the total pollution quota, Q*. In Fig. 9.1, it was demonstrated that social efficiency would be obtained when marginal benefits of abatement are equated with the marginal damage costs inflicted by pollution. Thus, starting from this main premise, it ought to be clear that each firm should be charged to reduce its pollution up to the point where MAC of control at the firm level is equal to the marginal environmental cost it generates. We can achieve this state of affairs either of the two ways: (1) a (*Pigouvian*) tax/subsidy can be imposed upon the firms to induce them to reach the golden rule of optimality stated above, or (2) a binding quota can be imposed on each firm such that while for each firm the respective optimality conditions are satisfied, the sum of the quantities of pollution generated by each firm adds up to Q*.

Both instruments should end up to satisfy the requirement that the respective marginal costs of abatement for both firms ought to be equated: $MAC_A = MAC_B$. Since these MAC curves are increasing with the quantity of pollution reduced, we can make use of a diagram with two vertical axes and where the size of the

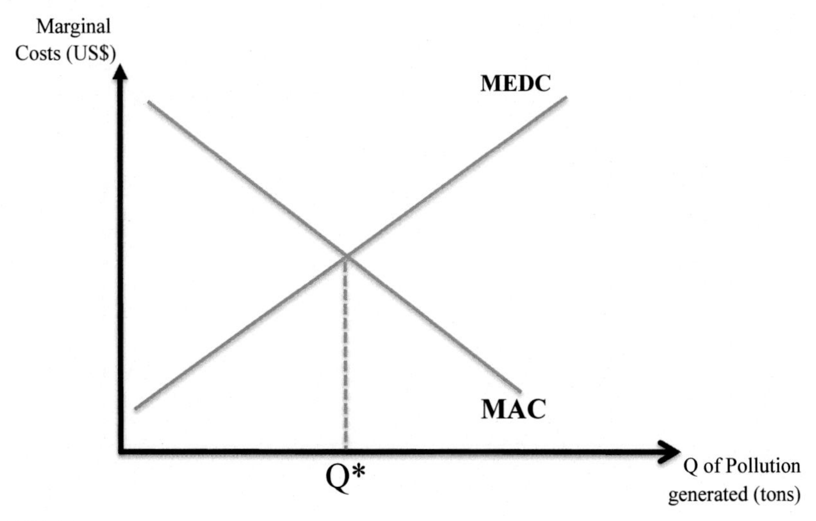

FIGURE 9.1

Social efficiency in pollution abatement.

horizontal axis is Q*. Suppose that Q* is taken as 100, and we want to achieve the optimal allocation rule at $MAC_A = MAC_B$ in Fig. 9.2.

Suppose as in Fig. 9.2, firm A is more "efficient" in controlling its emissions, so that its MAC curve is "flatter." To see the workings of a *cap-and-trade* system to allow for the market to allocate these 100 units of pollution, suppose that the authorities simply allocate (by auction pricing or simply by handing out) 50 units of allowances (rights to pollute) each. Thus, to begin with each firm has 50 units of tradable *rights-to-pollute* and 50 units of *uncontrolled* emissions.

In Fig. 9.2, this initial allocation is spotted as point X for firm B, and point Y for firm A. There is a clear incentive to trade. The marginal cost of abatement for firm B is higher than that of firm A, and thus firm A can sell allowances to B and B can reduce the amount of emissions it has to control (i.e., may expand its pollution level) by going back to its origin O_B.

Market trading will continue until the arbitrage opportunities are exhausted completely at point Z. Here, starting from the initial arbitrary allocation, firm A would have sold 20 units of *rights-to-pollute* allowances and cover 70 of the total 100 units of emissions to be controlled, whereas firm B purchases these 20 to expand its pollution level while reducing its share of controlled pollution obligations to 30 units. Total allocations remain 100 as desired.

The markets' *cap-and-trade* capability can be substituted or complemented by a taxation system, which, by trial and error, can enforce the same level of emissions abatement. Yet, many social and institutional problems await us. Now, let us turn to the real-world phenomena surrounding the abstractions of this optimality game.

Policies toward pricing carbon

The main principle of the policies destined for *pricing carbon* is to put a market value per ton of CO_2 emitted. The underlying motive is to change the behaviors

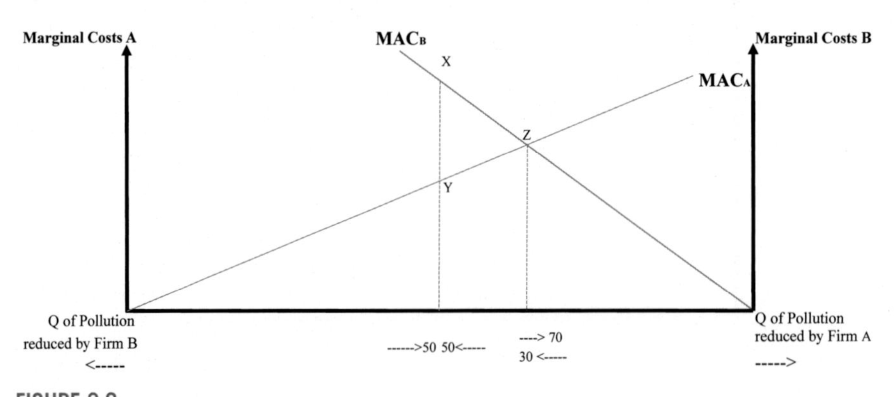

FIGURE 9.2

Trading for "right-to-pollute."

of the market agents by way of increasing the costs of releasing carbon as a result of their economic decisions. Given these incentives, it is expected that the market participants would internalize the costs of an otherwise unpriced environmental damage that had been incurred.

We will basically note two main sets of instruments under this heading: applying a direct carbon tax (per ton of CO_2 emitted) and generating an independent market for the sale and purchase of *right-to-pollute* emissions certificates.

Carbon taxation

The carbon tax is applied on ton of carbon that had been emitted. Ideally, by applying a single tax rate associated with the release of carbon, the policy aims at achieving *a second best solution* in the administration of the tax regime. Starting with the Swedish experience as early as 1991, now more than 40 national economies and federal states chose to administer one way or another a carbon taxation regime.

On the other hand, by an indirect way, many other forms of energy taxes on coal, natural gas, and other sources of energy inputs have consequences on mitigation of carbon emissions. According to OECD (2015), taxes on energy comprise roughly two-thirds of total environmental taxes applied.

The most significant advantage of carbon taxation systems is its institutional capability. Taxes are to be applied on the already established fiscal framework, with, hopefully, years of administrative experience of the government bureaucracy. Experience narrates that countries often prefer to increase the tax rate proportionally over the years. In Sweden, the tax rate had been raised from US$133 in 1991 to US$160 by 2013 under the Green Reform Programme. Likewise in Ireland, the tax rate had been increased from 15 to 20 euros per ton, whereas in Switzerland, the tax rate has been increased from CHF36 in 2014 to CHF84 in 2016 to be able to meet the 2020 emission targets (World Bank, 2015; OECD, 2015).

In certain others, coverage of the tax base had been expanded across sectors. In Finland, for instance, carbon taxation has been extended over to heating and transportation after it had originally been initiated over the electricity sector. The administrative body was also discussed to cover other greenhouse gases (GHGs) besides carbon dioxide, such as methane emissions from the processes associated with natural gas production (UNFCCC, US NDC, 2015).

Emission trading systems

The second approach works from the "quantity" side and sets a cap on aggregate emissions allowed. The market participants then trade among themselves the amount of allowances and settle at a price that clears the market for carbon units, typically tons. Thus, we set three steps: (1) determine the aggregate level (quota) of carbon emissions to be targeted; (2) allocate this total available quota among the participants by either auctioning the "allowances" or simply distributing them at some ad hoc prior rule; and finally (3) set up an institutional/legal framework wherein the market participants trade in those certificates to optimize on their market activities and decisions. Given its rather market-friendly stance, the ETS is regarded as more of a politically acceptable intervention.

The other advantage of the ETS is that it (at least on theoretical grounds) confirms the aggregate level of emissions, rather than achieving the same level of abatement by trial and error depending on the tax elasticity of the price system. This certainty has definite appeal. Yet, to be able to be operational, total amount of the carbon quota has to be severely binding. Otherwise, if all agents are comfortable with the initial levels of allowances allocated to them, then there would be no reason to engage into trade. This had been one of the main obstacles of the current implementations of the ETS across Europe (UNEP, 2015; OECD, 2013).

This across the board upper limit on maximum allowances determines both the size and efficiency of the market.

As stated above, the *cap-and-trade* ETS sounds as a more efficient and market-friendly mechanism from the point of view of economic theory. Yet in real-life practice, there remain many unresolved issues and critiques of the system. One concern is that as the ETS maintains an upper limit on aggregate emissions and runs its own course of action, it tends to preclude the applicability and efficiency of alternative instruments of control. Problems associated with control and monitoring are also a matter of administrative concern. To ensure effective controllability, it became necessary in most instances to complement the national monitoring systems with internationally accredited institutions, and all of this increased the costs of surveillance and generated a rent-seeking oriented bureaucracy, contrary to expectations of market efficiency otherwise. For instance, the expulsion of Greece and Bulgaria from the carbon trading systems because of their negligence to abide with the UN GHG reports is one evidence to this end.[1]

Another criticism toward the ETS is the concern for, what is known as, *the carbon leakage* phenomena. Due to impartiality of the mechanism across geographies, strict control and regulation in advanced industrialized economies thus tend to divert polluting production activities toward the sites where environmental standards are lax and uneven. This, of course, is part of a coordination problem across the globe and does not necessarily detriment the theoretical infrastructure of the ETS. Nevertheless, the evidence accumulated thus far on the *real politik* of the carbon trading system; its administrative promises of successful application have not been yet satisfactory.

Finally, a major problem associated with the design of an ETS is the start of the very mechanism: *how to allocate the allowances initially*? Past, and yet limited, experience thus far had been that of allocating the initial allowances (to right to pollute) individually based on past observations on emissions (which by itself had been accused of being a troublesome and croony system); creation of an effective "market" could not have been possible. As a result, the "market clearing price" of

[1] http://www.reuters.com/article/us-bulgaria-co2-suspension-idUSTRE65S3RU20100629.

carbon has been observed to be excessively volatile, often falling all the way down to its zero bound, and proven to be ineffective in curtailing emissions in any effective way.

Several proposals have been to remedy this. One is to impose a "*price corridor*," such that the market price is restricted to remain within. This will enable the market participants to form clearer expectations and guide the market toward a "positive" optimal price of carbon. A second complementary proposal is to establish a "reserve system" to maintain a positive and stable price. The European Union plans to establish such a mechanism as of 2019. Finally, a third, and again complementary, approach might be to generate viable *supply-side* measures. These may take the form of more effective auctioning mechanisms, elimination of frictions and carbon leakage, expanding the information set for participants, so on.

Needles to underline, all such interventions to price something "*not-pricable*" such as carbon emissions are not perfect, and the efficiency considerations based on an *imaginary first best optimality world* are subject to failed expectations. Now let us dwell on some of the nonmarket instruments that would aid abatement.

Regulations toward environmentally friendly economics

On the other side of the coin are possible regulatory mechanisms and nonmarket institutions to induce economic agents to demand for more environmental products. We take these in turn.

Renewables

Shifting to renewable sources of energy and production methods, in general, are increasingly regarded as a viable option not only to generate a decarbonized production system but also to obtain a more sustained economic path of development. Renewables have the direct advantage of achieving decarbonization and also opening new opportunities of investment and employment. Reducing emissions from burning of fossil fuels; eliminating risks involved during the production, storage, transportation, and consumption of hazardous energy inputs; and also reducing import dependency on energy sources are all regarded as important positives of securing an energy network based on renewables.

"Renewable energy certificates" (RECs) are also viable market-friendly instruments to attain these objectives. The REC system aims at creating incentives toward demanding "green" energy. An REC is generally designed so as to inform the end user that the source of energy (stated typically in megawatt hours of electricity) had been produced from green methods. The system has been popularized in the United States under the Clinton administration and now has become part of the traditional kit in countries such as India, Brazil, and Republic of China.

A significant prior condition for the REC system to serve as an effective regulation mechanism and to become a sustained financial instrument is pertinent that it is supported via a reliable system of effective monitoring and credible accounting. Many countries try to achieve this by way of releasing their reports to be audited by the international accreditation agencies.

On a more general level, one of the main concerns inhibiting the spread of *renewable* sources of energy is its cost structure and uncertainties involved in production. The levelized costs of energy are still a matter of debate, given the volatility of the markets and also the price distortions as a result of the existing generous subsidies in fossil fuels. Subsidies to coal production and consumption, in particular, disguise the true costs of fossils and distort incentives (see, e.g., Bridle and Kitson, 2017).

Other instruments

Many other innovations have been introduced. Fiscal incentives toward adopting renewable technologies, both at the R&D and at the ultimate production end, had been envisaged with different degrees of success. Such subsidies have also taken the form of creating incentives to creating and sustaining a market for renewable energy certificates and green finance.

Some forms of subsidies have taken the form of delivering a level of targeted quantity and therefore are termed as "quantity-based subsidies." Here the authority sets a requirement to produce a given level of electricity to be generated from renewable methods and offers subsidization of such targets. Here the price of unit electricity continues to be determined in the market. Australia's "renewable energy targets," India's "renewable energy standards," the European Union's "Tradable Green Certificates," and the United States' "Renewable Portfolio Standards" are examples of such policy framework on quantity-based subsidy administration.

Commensurate with above, some governments complemented above instruments with a focus on directing the market price. Here, a target price, often below the otherwise free market price, is declared, and the difference is paid by the fiscal authority. This may also take the form of purchase agreements at a prior set price guarantee. The price guarantee is typically set higher in industries based on solar, wind, and hydro sources. The International Energy Agency reports, for instance, that currently about 70 countries have been implementing price guaranteed purchase agreements, 40 of which are developing economies.

Overall assessments and concluding comments

As we have seen, all market instruments destined for abatement work, in a nutshell, through the cost minimization principle to divert away from polluting economic activities to greener ones. As almost all of these instruments entail, one way or another, an intervention to the price mechanism, the resulting resource allocation will necessary be of the *second best* in nature, calculated in arithmetics of static efficiency.

Nevertheless, many grim realities add up to our calculations. The IEA reports indicate that since the dawn of the industrial revolution (roughly 250 years) a cumulative total of 1650 Gt of CO_2 equivalent GHGs had been emitted to our planet's atmosphere. What is alarming is that more than half (833 Gt) had been emitted in the past 30 years since 1988. This acceleration, if continues, will result in an increase of at least 7°C of warming in our planet's surface temperature with commensurate threats to our livelihood on earth.

What complicates the matters is that in all our analyses above, *the unit of economic entity* has been typically regarded as "nations." This, however, is a distorted characterization of the global production and marketing stage of the global economy. Most of the largest producers of manufacturing and food industries in the world today are not "nations" but *transnational enterprises*. Their activities expand over national boundaries and enable them to find pollution havens where environmental standards are not applied strictly. The *2017 Carbon Majors Report* of the nonprofit *CDP*, for instance, document that 25 corporate and state-owned transnationals account for 51% of global GHG emissions. Add this number to reach the 100 largest corporate production units, and we reach 72% of the global industrial GHG emissions (CDP, 2017). CDP's report further reveals that "of the 635 Gt CO_2 equivalent green house gas emissions of the 100 active fossil fuel producers, 32% is public investor-owned, 9% is private investor owned and 59% is state-owned." Accordingly, the highest emitting companies since 1988 that are investor-owned include companies such as BP, Chevron, ExxonMobil, Shell, Total, Peabody, and BHP Billiton (CDP, 2017).

The size of these companies and the fact their activities range from energy production all the way to downstream marketing services complicate the effectiveness of abatement instruments. All this resounds in David Harvey's remark that "if a true price for carbon would in fact be implemented to cover its social costs, capitalism would not be able to operate." Given this remark, one of the key questions is "do the currently advanced economies still need net positive rate of economic growth?" Can't we envisage a steady-state rate of per capita real growth compatible with zero fossil fuel usage, zero net depletion of environmental resources at constant profit and wage shares?

References

Bridle, R., Kitson, L., 2017. Fossil-fuel Subsidies and Renewable Energy. IISD, Winnipeg.

CDP, 2017. 2017 Carbon Majors Report. London.

OECD, 2015. Divestment and stranded assets in the low-carbon transition. Background Paper for the 32nd Round Table on Sustainable Development.

OECD, 2013. Climate and carbon, aligning pieces and policies. OECD Environment Policy Paper (1).

UNEP, 2015. The Emissions Gap Report 2015: A UNEP Synthesis Report. http://uneplive.unep.org/media/docs/theme/13/EGR_2015_301115_lores.pdf.

World Bank, 2015. State and Trends of Carbon Pricing (Washington DC).

Further reading

OECD/IEA/NEA/ITF, 2015. Aligning Policies for a Low-Carbon Economy. OECD.

UNFCCC, 2015b. Synthesis Report on the Aggregate Effect of the Intended Nationally Determined Contributions, FCCC/CP/2015/7. http://unfccc.int/resource/docs/2015/cop21/eng/07.pdf.

CHAPTER

Financing the green economy

10

Burcu Ünüvar

Green economy to build green finance

The concept of green economy is only three decades old. David Pearce, Anil Markandya, and Ed Barbier are deemed to be the firsts to spell it loud in a report called "Blueprint for a Green Economy" published by London Environmental Economics Center in 1989.

The UN Environment defines green economy as "an economy that results in improved human well-being and social equity, while significantly reducing environmental risks and ecological scarcities."[1]

Following the aforementioned definition, United Nations Economic Commission for Europe (UNECE) points at different aspects of green economy, including but not limited to[2] environmental, social, and economic aspects. This multidimensional nature of green economy evidences that transition is not about greening certain selective areas, but it is about setting a new "green system" with its green markets, green institutions, green regulations, and green behaviors.

Greening the system via green finance

Transition to green economy requires linking economy, society, and environment. Thus a through transformation of production, consumption, and resource conservation plans should be rehandled with a green approach.

Obviously, aforementioned transformation ranges wide from spreading renewable energy sources, building sustainable production models to investing in resource efficient technologies, and conservation of resources. The Organisation for

[1] The Division for Sustainable Development Goals (DSDG) in the United Nations Department of Economic and Social Affairs: https://sustainabledevelopment.un.org/index.php?menu=1446 (last accessed on February 25, 2019).

[2] The United Nations Economic Commission for Europe (UNECE): https://www.unece.org/sustainable-development/green-economy/what-does-green-economy-mean.html (last accessed on February 25, 2019).

Handbook of Green Economics. https://doi.org/10.1016/B978-0-12-816635-2.00010-9

Economic Co-operation and Development (OECD) describes this transformation as the "biggest structural adjustment ever proposed in the field of international governance."[3]

No surprise, global community was quick to discover that a successful transition to green economy requires significant investment. Different sources estimate that some USD 90 trillion will be needed in the coming two decades to achieve global sustainable development and climate objectives. Hence the transition to green economy should go hand in hand with the development of green finance, their success being interdependent.

Although the need for green finance is generally accepted, there is yet no consensus around the definition of green finance. G20 Green Finance Study Group launched during China's presidency of G20 in 2016 defines green finance as the "financing of the investments that provide environmental benefits in the broader context of sustainable development."

Asian Development Bank writes "green finance involves engaging traditional capital markets in creating and distributing a range of financial products and services that deliver both investable returns and environmentally positive outcomes."[4]

Lindenberg (2014) proposes a definition of green finance that is based on three main pillars[5]:

1. The financing of public and private green investments (including preparatory and capital costs) in the following areas:
   a. Environmental goods and services (such as water management and protection of biodiversity and landscapes).
   b. Prevention, minimization, and compensation of damages to the environment and to the climate (such as energy efficiency or dams).
2. The financing of public policies (including operational costs) that encourage the implementation of environmental and environmental damage mitigation or adaptation projects and initiatives (for example, feed-in tariffs for renewable energies).
3. Components of the financial system that deal specifically with green investments, such as the Green Climate Fund or financial instruments for green investments (e.g., green bonds and structured green funds), including their specific legal, economic, and institutional framework conditions.

Defining green finance is an important issue, as measurement and planning become vague when there are no boundaries. Although there is no strict and

---

[3] OECD: http://www.oecd.org/cgfi/about/ (last accessed on February 25, 2019).

[4] Asian Development Bank: https://development.asia/explainer/green-finance-explained (last accessed on February 25, 2019).

[5] https://www.die-gdi.de/uploads/media/Lindenberg_Definition_green_finance.pdf (last accessed on February 25, 2019).

clear-cut borders, studies in hand are developed enough to show us where to look at while defining green financial activities.

Maheshwari, Avendano, Stein (2016)[6] ran a survey across financial institutions on the sectors/activities that they include in their definition of green finance. The International Finance Corporation (IFC) reports the findings of the aforementioned study. Accordingly, the respondents prioritized the following broad categories[7]:

- Adaptation (conservation, biosystem adaptation)
- Carbon capture and storage
- Energy efficiency (cogeneration, smart grid)
- Environmental protection (pollution control, prevention, and treatment)
- Green buildings
- Green products and materials
- Renewable energy (solar, wind, hydro)
- Sustainable land management (sustainable agriculture, forestry)
- Transport (urban rail/metro, electric, hybrid)
- Waste management (recycling, waste management)
- Water (water efficiency, wastewater treatment)

Some sectors such as renewable energy and green buildings are easy to have a consensus around, in terms of being green and/or linking to green finance. Yet, this is not the case for other areas such as nuclear power and noise abatement.

Clearly, government, financial institutions, and international organizations have a tendency to define green finance in a manner that stands closest to their own business strategies. Therefore, despite rising number of international organizations working for the harmonization of green finance activities/instruments, standardization still stands as a problem.

Meanwhile, having some "relative" areas to green finance such as climate finance, sustainable finance, and environmental finance also widens the area and makes it difficult to track the "green data." Despite this difficulty, embracing these areas is of crucial importance, as sticking to a strict definition risks excluding some participants. Therefore, sustainable finance and climate finance have been gaining ground as derivatives of "green logic" and accepted as means of enlarging the impact area.

[6] Measuring Progress on Green Finance—Findings from a Survey, Maheshwari, Avendano, Stein: http://unepinquiry.org/wp-content/uploads/2016/09/5_Outline_Framework_for_Measuring_Progress_on_Green_Finance.pdf.

[7] International Finance Corporation: https://www.ifc.org/wps/wcm/connect/48d24e3b-2e37-4539-8a5e-a8b4d6e6acac/IFC_Green+Finance+-+A+Bottom-up+Approach+to+Track+Existing+Flows+2017.pdf?MOD=AJPERES (last accessed on February 25, 2019).

Who are the green actors?

During the very early stages of building green finance, people were in search of a leading actor. Now it became more or less consensus that this is not a one-entity show and we need all concerned parties around the table, with different but complementary contributions.

Experience to date showed that both public and private sectors should be active in the process of building and spreading green finance approach. Central banks and regulatory authorities have the power to push the market dynamics toward a greener path and make sure that the path is followed diligently. Finally yet importantly, development banks and financial institutions are the sine qua non in this process for expanding the scope of the green financial activities.

Consensus does not deny the contribution of the government support at the infant stage of green finance. Public's role is usually defined through the means of incentives and regulations. Yet facilitating to shape the public opinion should not be undermined. It has an undeniable pressure over the private sector to "go green." Peng et al. (2018) argue that government should deliver stimulus policies while also providing support from philosophy to policy on green investment.

Public sector promoting transparency and consistency in the green reporting is vital. Without such an effort, tracking green finance cannot be accomplished, and the system will be more vulnerable to abuses.

Insisting on "greening the system beyond green instruments" is easier said than done. Private sector is the most likely pioneering actor in this model to integrate green decision-making into the logic of income statement, i.e., planning green revenues and green expenditures.

Specific to green finance, reporting of the "use of proceeds" is crucial. This requires private sector adapting a transparent reporting aligned with environmental, social, and governance (ESG) criteria. While still being at the development stage, "blended finance," i.e., cooperation of public and private financing facilities within the same project is crucial to further feed the growth potential of green finance. This also enables public and private sector feeding each other through their experiences.

Development banks and multinational organizations, too, play catalyzing roles to further facilitate tracking standards. But contributions of these organizations will weigh more through their widespread sectoral and country experiences. It will be only through the contribution of the development banks and multinationals that bottom-up needs of the different sectors and countries at different stages of development cycle can meet top-down green solutions.

Central banks are also a part of the going green debate. Climate risks and related factors have the potential to affect price stability as well as the financial stability, pulling the central bank into the green finance story. Dikau and Volz (2018)[8] position

[8] https://eprints.soas.ac.uk/26445/1/Dikau%20Volz%202018%20Central%20Banking,%20Climate%20Change%20and%20Green%20Finance.pdf (last accessed on February 25, 2019).

the roles of central banks in green finance through monetary policy tools including but not limited to green microprudential and macroprudential regulations.

Why should investors worry about climate risks?

Markets are blamed to be "myopic" but not deaf or unaware. Risks associated with the climate change are multidimensional and have obvious short- to medium-term repercussions that should be taken into consideration by the financial market players. One can broadly classify the economic impacts of climate change according to their fiscal and monetary aspects (Fig. 10.1).

Pressure over the budget

Building resilience against shocks related with extreme weather events requires significant investment mainly into energy, construction, and transport sectors. Natural disasters are hard to predict (if not impossible), thus delivering a hard-to-foresee cleaning and rebuilding cost in the aftermath of the disaster.

Higher health-related spending

Climate-related risks threaten the public health. It does not only elevate the health expenditures but also hurt the human capital of the society.

Many tropical zone countries are poor and more susceptible to climate risks. Meanwhile, since lowest income groups of the society usually have the weakest adaptation skills, climate-related risks have uneven impacts, deepening the current inequality.

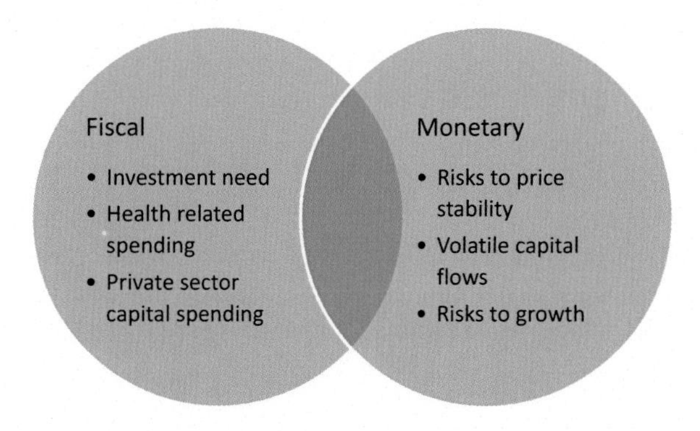

FIGURE 10.1

Some climate-related factors that can affect investments.

Source: Author's Own Illustration.

Disruption in private sector capital spending

Any breaking down in the supply chain due to climate-related factors imposes further pressure to the production line, thus creating unforeseen investment need for the private sector and leading to disruption.

Risks to price stability

The World Economic Forum reports that global crop prices will be 20% higher in 2050 than they would have been without climate change.[9] Higher temperature and shorter growing seasons are expected to elevate food prices while also making them more volatile. Meanwhile, energy prices also cast shadow over the pricing dynamics.

Volatile capital flows

As climate-related factors started to weigh more in the operational strategy of the multinational and international companies, regions that are more prone to extreme climate risk factors will be disadvantageous in attracting foreign capital.

Pressure over the growth performance

Damage by extreme weather conditions are expensive to handle and create a volatile short-term growth path. The United Nations Office for Disaster Risk Reduction (UNISDR) and the Center for Research on the Epidemiology of Disasters (CRED) report that direct economic losses from extreme weather events rose by 151% between 1978–97 period and 1998–2017 period.[10]

Depleting the natural resources jeopardizes the potential growth of a country. Damage from gradual global warming hurts the medium-term growth dynamics permanently.

Environmental commodities

Obviously, there are good reasons to take the green factors into consideration for both today and tomorrow. Green finance instruments step right here to tame, plan, and administer/manage.

Environmental commodities can be deemed as nontangible energy credits. The value of the commodity is determined by the need of the purchaser to generate and consume clean energy. Main motivation that triggered trading of environmental

[9] http://reports.weforum.org/global-risks-2016/climate-change-and-risks-to-food-security/#view/fn-16 (last accessed on February 25, 2019).

[10] https://www.unisdr.org/files/61119_credeconomiclosses.pdf (last accessed on February 25, 2019).

commodities was to promote clean energy production/consumption while also tackling greenhouse emissions.

As cap over the pollution became tighter, it started to make sense to pay for the right to pollute, which fuels the continuation of the carbon trading markets.

Henriquez (2013) notes that "market-based solutions to environmental problems offer great promise but require complex public policies that take into account the many institutional factors necessary for the market to work and that guard against the social forces that can derail good public policies." Indeed, public front most often has two approaches to build the environmental commodity trading system: penalizing polluters and rewarding clean energy producers.

Kyoto Protocol is a landmark in the development of the environmental commodities market. By delivering internationally binding emission reduction targets, the Protocol paved the way for carbon emission trading and related financial instruments. Leading instruments in the market can be listed as follows:

1. Carbon pricing:
   The World Bank defines carbon pricing[11] as "an instrument that captures the external costs of greenhouse gas emissions and ties them to their sources through a price, usually in the form of a price on the carbon dioxide emitted." The logic is pretty straightforward in the way that creating a cost for carbon emission serves to push the environmental cost back to the one that created the initial carbon burden. The system sends financial signals enabling polluters to decide for themselves with the options of stopping/reducing polluting, reduce emissions, or continue to pollute but bearing the cost.
   Emission trading systems (ETSs) and carbon taxes are usually the most known carbon trading mechanisms.
   a. Emission trading systems:
      i. Cap and trade: The system puts a cap over the gas emissions and enables those who stand below the cap to sell their "spare right" to those who exceeds their caps. This system creates a supply and demand mechanism that paves the way for pricing of the emission.
      ii. Baseline and credit: Baseline emission levels are determined for individual entities, and those entities that could lower their emissions are granted with credits, which can be sold to those that exceed their limit.
   b. Carbon tax: It is a rather more direct instrument in the sense that it sets a price on carbon by defining a tax rate on greenhouse emissions or the carbon content of the fuels.
2. Carbon offset mechanism:
   Climate mitigation debates triggered the emergence of carbon offset mechanism. It differs from the cap-and-trade in the sense that its mechanism is outside the

[11] https://carbonpricingdashboard.worldbank.org/what-carbon-pricing (last accessed on February 25, 2019).

mandated caps, i.e., they reward clean energy production with the means that can be outside company's main operations.

3. Internal carbon pricing:
   Companies internally design this system to estimate the cost of carbon emissions. These estimates further guide the investment decisions for energy efficiency.
4. Renewable energy certificates (RECs):
   They are tradable energy certificates, based on the proof of generation of energy from renewable energy sources with contribution to the shared power system.
5. White certificates:
   They are some types of RECs that are devoted to energy efficiency gains rather than the production of renewable energy.
   According to the World Bank's "State and Trends of Carbon Pricing (2018)" Report dating May 2018[12] (covers until April 1, 2018):
   - 45 national and 25 subnational jurisdictions are putting a price on carbon.
   - Carbon pricing initiatives implemented and scheduled for implementation about 20% of global green house emissions, up from 15% in 2017.
   - Governments raised USD 33 billion in carbon pricing revenues in 2017 compared with USD 22 billion raised in 2016.
   - In 2018, the total value of ETSs and carbon taxes is USD 82 billion, representing a 56% YoY increase (Box 10.1).

Box 10.1 The task force on climate-related financial disclosures

The Task Force on Climate-related Financial Disclosures (TCFD) was established on December 4, 2015. The TCFD "seeks to develop recommendations for voluntary climate-related financial disclosures that are consistent, comparable, reliable, clear, and efficient and provide decision-useful information to lenders, insurers, and investors."[13]

Climate-related financial disclosures are based on metrics and targets to assess climate-related risks as well as the opportunities. Such an approach delivers the risk management stance that lies in the core of the TCFD logic, paving the way for building climate-aware strategies as well as improving the governance.

As climate-related financial disclosure evolves, carbon pricing becomes a popular metric to integrate climate-related risks. The World Bank reports that a growing number of organizations use internal carbon pricing as a tool to mitigate climate-related financial risks while also planning a transition to low-carbon economy.

[12] https://openknowledge.worldbank.org/bitstream/handle/10986/29687/9781464812927.pdf (last accessed on February 25, 2019).

[13] https://www.fsb-tcfd.org/about/# (last accessed on February 25, 2019).

Green financial instruments

Green financial products have been in the market for long. The Green Finance Initiative notes that emission trading was first considered in the 1960s. Yet, creation of the green finance sector as we understand it today is much younger. Many people mark the issuance of first green bond in 2007 as the beginning of the conventional green finance sector.

Green finance is mainly associated with the green bonds at a glance, but there are many other products[14]:

- Green loans
- Carbon tilted indices
- Municipal green project financing
- Green crowdfunding platform
- Green investment banks
- Renewable yieldcos
- Catastrophe bonds
- Green insurance
- Green funds

Green bonds

The World Bank describes green bond as "plain vanilla fixed income product that offers investors the opportunity to participate in the financing of green projects that help mitigate climate change and help countries adapt to the effects of climate change."[15]

The Climate Bonds Initiative (CBI), an international investor-focused not-for-profit organization, marks 2007 as the year green bond market kicks off with the AAA-rated issuance of the European Investment Bank and the World Bank. But wider action in the green bond market started in 2013. The first municipal green bond issue also came in 2013 to be followed by the first corporate green bond issue that came in 2014.

There are different types of green bonds, depending on the use of proceeds or project earmarkings. A green bond might differ in the way it is earmarked for a project or pool of projects, thus having different names such "Use of proceeds bonds," "Project bonds," or "revenue bonds".

Some advantages of green bonds

- For issuer:
    - It can attract capital for low-carbon assets

[14] http://greenfinanceinitiative.org/facts-figures/ (Last accessed on February 25, 2019).

[15] http://worldbank.or.jp/debtsecurities/web/Euromoney_2010_Handbook_Environmental_Finance.pdf (last accessed on February 25, 2019).

  - Signaling green/environmental/sustainable commitment of the issuer
  - Potential access to wider investor base
- For investors:
  - Transparent reporting
  - Hedge against environmental, carbon-related risks
  - Improves the ESG profile of the investor

Headline figures from the green bond market

According to the "State of the Market" report prepared by the CBI (covering 2005-1H2018):

- There are 498 green bond issuers with USD 389 billion of outstanding bond value.
- Labeled green bonds with energy allocation represent a third of the outstanding sector amount (USD 90 billion).
- Green bonds represent 17% of outstanding issuance in the water theme.

As the green bonds become more popular, exchange representation has also developed. Table 10.1 exhibits some stock exchanges that have green bond segments, supporting the liquidity of the instrument while also promoting the green finance.

Green bonds principles

In 2014, the International Capital Market Association (ICMA) launched voluntary process guidelines for issuing green bonds called Green Bonds Principles (GBP). Principles aim at "promoting the integrity in the Green Bond market through guidelines that recommend transparency, disclosure, and reporting" and being updated with intervals.[16]

GBP has four core components:

1. Use of proceeds:
   Since the proceeds from the green bonds will exclusively finance or refinance eligible green projects, this principle lies in the heart of the list. Accordingly, all designated projects should provide clear environmental benefits that can be assessed and quantified.
   There are several broad categories recognized by the principles: climate change mitigation, climate change adaptation, natural resource conservation, biodiversity conservation, pollution prevention and control, etc.

[16] Green Bond Principles—Voluntary Process Guidelines for Issuing Green Bonds-2018: https://www.icmagroup.org/green-social-and-sustainability-bonds/green-bond-principles-gbp/ (last accessed on February 25, 2019).

Table 10.1 Green bond segments on some stock exchanges.

Name of stock exchange
Oslo Stock Exchange
London Stock Exchange
Nasdaq Sustainable Bond Market
Luxemburg Stock Exchange
Johannesburg Stock Exchange
Japan Exchange Group

Source: Author's Own Search from the Original Web Sources: https://www.oslobors.no/ob_eng/Oslo-Boers/Listing/Interest-bearing-instruments/Green-bonds, https://www.lseg.com/sustainable, https://business.nasdaq.com/list/listing-options/European-Markets/nordic-fixed-income/sustainable-bonds, https://www.bourse.lu/luxse, https://www.jse.co.za/articles/Pages/JSE-launches-Green-Bond-segment-to-fund-low-carbon-projects.aspx, https://www.jpx.co.jp/english/equities/products/tpbm/green-and-social-bonds/index.html.

2. Process for project evaluation and selection:
   GBP asks the issuer to clearly communicate to investors:
   a. The environmental sustainability objectives
   b. The process evidencing how the particular project fits eligibility
   c. The related eligibility criteria
   High level of transparency is encouraged to be supplemented by an external review.
3. Management of proceeds:
   The proceeds of the green bond or an equal amount should be credited to an account that can be tracked. Until the bond matures, the balance of the account should be adjusted. Meanwhile, it is highly recommended to have an auditor or a third party opinion throughout this process.
4. Reporting:
   Issuers are encouraged to submit regular, readily available, transparent reporting showing their methodology, assumptions, and key data.
   These volunteer guidelines also aim at preventing "greenwashing," in which the seller either promises unrealistically high environmental benefits or covers the environmental damage. Yet there is still a long way to walk in terms of monitoring and verification of data and reporting.

Green loans

The Loan Market Association (LMA) defines green loans as "any type of loan instrument made available exclusively to finance or refinance, in whole or in part, new and/or existing eligible green projects."[17]

[17] Green Loans Principles—December 2018 https://www.lma.eu.com/application/files/9115/4452/5458/741_LM_Green_Loan_Principles_Booklet_V8.pdf (last accessed on February 25, 2019).

Despite showing significant discrepancy in size among countries, the use of green loans is getting popular in many different countries,[18] with outstanding loan amounts varying from USD 56.8 billion in the United States and USD 13 billion in the United Kingdom to USD 1.4 billion in Italy and USD 1.3 billion in Ireland.

Green loans vary also in type, some of which are as follows:

1. Green project finance: Cash flows generated by a project/portfolio of projects serve as the collateral.
2. Green syndicated loans: Lending facility will be rendered by a "group of banks."
3. Green bilateral loans: Borrower and lender have corporate guarantee in between.

Green loans principles

At the beginning of their launch, green loans were also subject to GBP. Green Loans Principles emerged early 2018 to promote the development and integrity of the market.

Green Loans Principles aims at setting a framework that can be understood easily by all market participants for clarifying the nature of green loans following four pillars:

- Use of proceeds: All designated green projects linked to the loan should have clear environmental benefits, which can be assessed, and where feasible, quantified, measured, and reported by the borrower. The principle recognizes that definitions of green and green projects may vary depending on sector and geography.
- Process of project evaluation and selection: The environmental sustainability objectives, how the project fits the eligibility, while documenting exclusion criteria, etc.
- Management of proceeds: Proceeds should be credited to a dedicated account or otherwise tracked by the borrower in an appropriate manner, maintaining transparency and promoting the integrity of the product.
- Reporting: A transparent reporting that includes a list of the green projects to which the green loan proceeds are to be allocated should be kept ready by the borrower.

Green sukuk

A sukuk is an Islamic bond that can generate returns to investors complying with the Sharia law. Accordingly, sukuk supplies a certificate, with proceeds used to purchase an asset that is mutually owned by buyer and seller. Proceeds from green sukuk must

---

[18] Green Finance: A Bottom Approach to Track Existing Flows, International Finance Corporation, 2017: https://www.ifc.org/wps/wcm/connect/48d24e3b-2e37-4539-8a5e-a8b4d6e6acac/IFC_Green-+Finance+-+A+Bottom-up+Approach+to+Track+Existing+Flows+2017.pdf?MOD=AJPERES (last accessed on February 25, 2019).

only be devoted to climate-friendly investments. It is open both for conventional and green investors, and the World Bank sees it "a big step toward bridging the gap between conventional financing and Islamic financing."[19]

Catastrophe bonds (cat bonds)

Catastrophe bonds are alternative to traditional insurance and reinsurance products.[20] Their coupon and principal payments depend on the nonoccurrence of a predefined catastrophic event, the performance of an insurance portfolio, or the value of an index of natural catastrophe risks.[21]

Climate bonds

Climate bonds are used to finance/refinance projects that address climate risks. The difference between "green bonds" and "climate bonds" takes significant attention. Climate bonds are represented by a broader set of bonds, whose proceeds are devoted to climate-related subjects but not necessarily labeled as green.

Indeed, according to the figures of CBI, climate-aligned bond universe reached USD 1.45 trillion of which USD 389 billion is labeled green bonds.

While tracking the issuance of climate bonds, CBI follows certain alignment criteria according to its taxonomy.

- Fully aligned climate issuers: Bond issuers that derive more than 95% of revenues from climate-aligned assets and green business lines.
- Strongly aligned climate issuers: Bond issuers where 75%–95% of revenues are derived from climate-aligned assets and green business lines (Graph 10.1).

Do the climate risks affect the cost of capital?

Developing countries usually run a high borrowing need on the back of young population, limited savings, population growth, and urbanization that fuel their need for investment. The dilemma for the developing countries stands right at this point where the investment need is high but so are the pressures over the cost of financing.

On a related front, in its World Economic Outlook Report dating October 2017, the International Monetary Fund-World Bank (IMF-WB) noted that "low-income developing countries and small states, respectively, are 5 and 200 times more likely to be hit by a weather-related natural disaster than the rest of the world." Meanwhile, the loss due to climate-related disasters have a bigger portion in the gross domestic product of the low-income countries.

[19] https://www.worldbank.org/en/news/infographic/2017/09/19/malaysia-green-sukuk (last accessed on February 25, 2019).
[20] Insurance Information Institute: https://www.iii.org/fact-statistic/facts-statistics-catastrophe-bonds (last accessed on February 25, 2019).
[21] https://www.climatebonds.net/cat-or-out (last accessed on February 25, 2019).

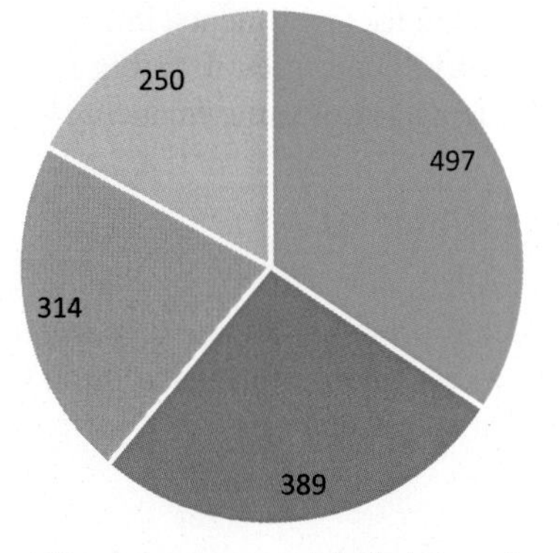

GRAPH 10.1

The climate-aligned bond universe, 1H2018, USD billion.

Source: Climate Bonds Initiative.

While the usually high borrowing cost of the low- and low- to middle-income countries is no surprise, studies showing the relationship between climate vulnerability and cost of capital are limited. However, considering the fiscal aspect of the climate-related risks, it should be most unusual to think that economic losses will not be translated into financial losses.

A report prepared by the Imperial College Business School and SOAS University of London notes the relationship between physical climate risks and country-level financial indicators.[22] Accordingly, weather shocks and/or climate trends have an impact on rating agencies' decision and on sovereign debt yields. Yet, there is a transmission mechanism in between going through economic and external indicators as well as fiscal and monetary indicators. Higher insurance premiums, loss of tax revenues, reduced economic output, stranded assets, social conflict, and trade imbalances are noted among these indicators.

In case of negative weather shocks or adverse climate trends, insurance premiums will be higher and/or high tax losses will come on to the table. Economic output will be under pressure, with trade imbalances and stranded assets, which might eventually trigger social conflict. Under such a scenario, it will be unrealistic to think that there will be no consequences on the financial markets front. Indeed, Graph 10.2 exhibits rising sovereign yields accompanied by high climate vulnerability.

[22] Climate Change and the Cost of Capital in Developing Countries: http://unepinquiry.org/wp-content/uploads/2018/07/Climate_Change_and_the_Cost_of_Capital_in_Developing_Countries.pdf.

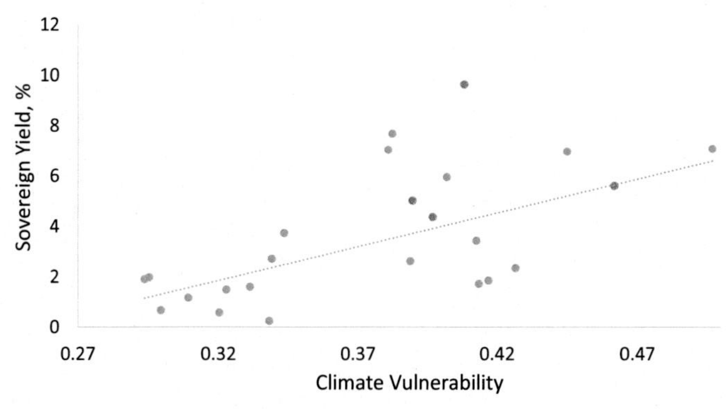

GRAPH 10.2

Climate vulnerability index and sovereign yields.
Red dots show the countries classified in Vulnerable-20 [Countries included into the study: Australia, Austria, Belgium, Brazil, Bulgari, Canada, Chile, China, Croatia, S. Cyprus, Czech Republic, Denmark, Finland, France, Germany, India, Indonesia, Ireland, Israel, Italy, Japan, Lithuania, Mexico, Netherlands, New Zealand, Peru, Poland, Portugal, Romania, Singapore, Slovakia, Slovenia, South Africa, Spain, Sweden, Switzerland, Thailand, the United Kingdom, the United States. Red dots are Colombia, Costa Rica, Lebanon, Philippines (V-20 members)].

Source: ND-Gain, Bloomberg, Author's Own Calculations, Data December 2018.

These consequences are getting more systematic now, as rating agencies started to include climate-related risks into their sovereign grading.

It is now generally accepted that climate-related risks have a share in sovereign risk profile. Credit rating agency Moody's is reported to be the first to mention the role of climate-related actions in risk profile as of 2016 to be followed by others.

While the story has been gaining popularity around the sovereign rating, climate-related issues affect corporate ratings as well.

To give an idea, S&P reviewed[23] the impact of environmental and climate risks and opportunities on their corporate credit ratings between mid-2015 and mid-2017.

Accordingly:

- There were 717 ratings where environmental and climate risks were important in the analysis.
- There were 106 ratings where environmental and climate risks were key to a rating action.
- Of these 106 cases, 44% of the actions were in the positive direction, whereas 56% of the actions were in the negative direction.

[23] S&P Global Ratings, COP24 Special Edition, Shining a Light on Climate Finance, December 2018: https://www.spglobal.com/en/research-insights/articles/easset_upload_file3707_847036_e.pdf (last accessed on February 25, 2019).

Just recently, Kling et al. (2018) assessed the impact of climate risks on sovereign borrowing costs. Authors worked with a sample of 46 countries made up of a selection of Vulnerable-20 (V-20) countries, the G-7, and a group of middle- to low-income countries not in V-20. The sample period is from 1996 to 2016.

Their work links climate vulnerability and social preparedness to cost of debt, leading to their primary conclusion that "countries with higher degrees of climate vulnerability face higher sovereign borrowing costs" confirming Graph 10.2. Climate vulnerability has an upward and significant impact on sovereign yields, whereas social preparedness has a downward and significant effect on bond yields.

Authors go into a more specific calculation for V20 countries. Calculating an average base cost of 12.4% for debt issues of V-20, their model predicts that climate vulnerability increases the cost of debt, on average by 117 basis points.

Can green finance fight high cost of capital?

Can green financing and particularly green bonds serve as panacea against high borrowing costs? The debate goes that issuers in the green bond universe can expect better pricing compared with a similar term vanilla product, explained by "green premium" or the so-called "greenium." There is no generally accepted answer to prove the existence of greenium (if there is at all). A usual approach for the issuer is to check if the new issue premium is smaller than it has been historically or lower than expected.

The CBI uses the term "greenium" to mean "green bonds that price inside their own yield curves." Schmitt (2017) finds a green bond premium of −3.2 bps yet points at supply and demand conditions for possible change in the yield.

The CBI and the International Finance Corporation released a report highlighting green bond pricing in the primary market in the first half of 2018.[24] The report built yield curves for 18 of 29 green bonds and ended up noting "no evidence of greenium." Yet findings about the secondary market performance are also interesting to note:

- In the immediate secondary market, 72% of green bonds had tighter spreads after 7 days.
- After 28 days, 62% of green bonds have tighter spreads.

Without betting on debatable greenium, there is more or less a consensus over certain positive side effects of green finance, which should be prioritized:

- The disclosure requirement that comes with the green issues usually curbs the information asymmetry.
- Due to the environmental screening, green bonds are deemed to have lower exposure to environmental risks. In fact, Preclaw and Bakshi (2015) suggest that if the standards of the green bonds are improved, they can be used as hedging mechanism against environmental risks.

[24] Green Bond Pricing in the Primary Market January-June 2018: https://www.climatebonds.net/files/reports/cbi_pricing_h1_2018_01l.pdf (last accessed on February 25, 2019).

Challenges and path ahead to unlock the green potential

Seeing the size of the financing need to reach the ambitious goals of Paris Climate Agreement, mobilizing capital stands as the main challenge. Yet, the high level of financing need makes it obvious that this can only be done by coordinated and complimentary actions of public and private sector as well as multinationals.

- Motivating and monitoring transparent reporting standards is crucial but not enough. Reporting should be designed in a way to support green thinking through a business perspective.
  - *Climate-related financial disclosures will help to embed the "green thinking" into an "income statement logic." Acting green will help to publish green financial statements, but committing to start publishing green financial statements will also encourage thinking/acting green.*
- Utilizing the fin-tech possibilities might help to make green supply and green demand meet, bridging temporal and geographical mismatches.
  - *The Green Digital Finance Alliance launched in 2017 stands as an important first practice.*
- Multilaterals seem to be the right address to pioneer bringing green projects to bankable stage in the emerging markets (EM) universe. Meanwhile, expanding the vague definition of "strictly green" to "sincerely green" in a transparent and quantifiable way and embracing sustainability perspective will help to enlarge the impact area of green finance. Such a controlled flexibility will fit the sectoral investment needs of the developing countries better. Development banks/ Development finance institutions (DFIs) will be the key in this strategy due to their in-depth country-specific experiences and wide sectoral coverage. Indeed, the Climate Policy Initiative notes that "national DFIs might be the next frontier, following the progress made by multilateral and bilateral DFIs on greening existing finance."[25]
  - *The Industrial Development Bank of Turkey (TSKB), a private development bank in Turkey, issued the first Green/Sustainable Bond of CEEMEA Region in 2016. The number of investors was more than double vis-à-vis the previous two conventional Eurobond issues of the Bank. The dual coverage of the bond, i.e., green-sustainable content, enabled to deliver financial solutions aligned with the sectoral landscape of the country. Such "first practices" are important to pave the way for bringing other issuers to the market.*
- Governments and policymakers can create smart and simple incentive systems to catalyze green finance:
  - *The Monetary Authority of Singapore covers the additional cost of issuing a green versus a vanilla bond. The Ministry of Environment in Japan*

[25] Global Landscape of Climate Finance (2017), Climate Policy Initiative: https://climatepolicyinitiative.org/wp-content/uploads/2017/10/2017-Global-Landscape-of-Climate-Finance.pdf (last accessed on February 25, 2019).

introduced subsidies of up to JPY 50 million per issue to help cover the costs of third party outsourcing such as external reviews and consultancy.[26]

- Central Banks should be alerted about the effects of climate-related risks on price stability and financial stability. Therefore, one can see increasing discussion developing around green macro- and microprudentials. Yet, there should be a thin red line while discussing the green central banking, i.e., not exceeding the role of monetary policy.
  - *The People's Bank of China (PBoC) will include green bonds as eligible collateral for its Medium-Term Lending Facility. Green credit is now part of PBoC's macroprudential assessment: the higher the volume of green assets held (loans, bonds, etc), the higher a bank's Macro-Prudential Assessment (MPA) score.*[27]

Concluding comments

Ambitious goals of the Paris Climate Agreement urge mobilization of capital in a way to meet the principles of green economy. Creating a consensus over a well-defined green economy-green finance link, embracing sustainability, environment, and climate is a good starting point while tackling the challenge.

Although the concept of green finance is relatively new, interest in the topic accelerated the growth in the market while also delivering a wider product basket. This fast pace of growth should be handled with care to avoid a premature and short-lived boost.

Transparent regulations and coordinated monitoring will be the key for a healthier functioning green financial system. All related parties, policymakers, private sector players, and development financial institutions should work as stakeholders at this development stage.

While creating financial depth for the green instruments will top the agenda, it should go hand in hand with enlarging the impact area of green financial logic. Thinking green finance beyond the products and urging stakeholders to "go green" in their activities, will help the means to justify the need.

Abbreviations

CBI	Climate Bonds Initiative
CRED	Center for Research on the Epidemiology of Disasters
DFI	Development finance institutions
ESG	Environmental, social, and governance

[26] Bonds and Climate Change, The State of the Market, 2018, Climate Bonds Initiative: https://www.climatebonds.net/files/reports/cbi_sotm_2018_final_01k-web.pdf.

[27] https://www.climatebonds.net/files/reports/cbi_sotm_2018_final_01k-web.pdf (last accessed on February 25, 2019).

GBP	Green Bonds Principles
ICMA	International Capital Market Association
IFC	International Finance Corporation
LMA	Loan Market Association
OECD	Organisation for Economic Co-operation and Development
TCFD	Task Force on Climate-related Financial Disclosure
UN	United Nations
UNECE	United Nations Economic Commission for Europe
UNISDR	United Nations Office for Disaster Risk Reduction

References

Dikau, S., Volz, U., 2018. Central Bank Mandates, Sustainability Objectives and the Promotion of Green Finance. mimeo, London: SOAS University of London.

Henríquez, B.L.P., 2013. Environmental Commodities Markets and Emissions Trading: Towards a Low-Carbon Future. Routledge.

Kling, G., Lo, Y.C., Murinde, V., Volz, U., 2018. Climate Vulnerability and the Cost of Debt.

Peng, H., Feng, T., Zhou, C., 2018. International experiences in the development of green finance. American Journal of Industrial and Business Management 8 (02), 385.

Preclaw, R., Bakshi, A., 2015. The Cost of Being Green. Report. Barclays Credit Research.

Schmitt, S., 2017. A Parametric Approach to Estimate the Green Bond Premium (Doctoral dissertation).

Index

'*Note:* Page numbers followed by "t" indicate tables, "f" indicate figures and "b" indicate boxes.'

G

H

I

T

U

W

Printed in the United States
By Bookmasters